Out of the Cradle

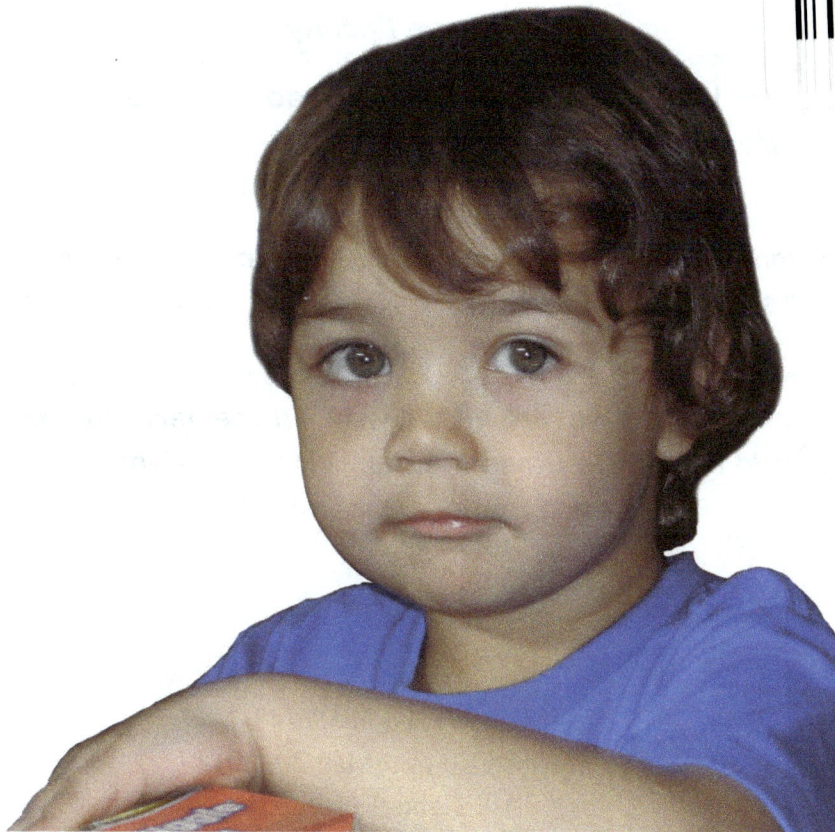

For all the grand kids

by Charles Lee Lesher

What They Are Saying

"Out of the Cradle is right on target! Congratulations on a good and important read!" Dr. David Williams, Director of NASA's Planetary Image Facility

◇◇◇

"Engaging! Well researched! Easy to read and a good review of the facts." Henry Geist, Physicist and Vice President of the Humanist Society of Greater Phoenix

◇◇◇

"A potent analysis of the energy challenges facing the world and their solutions." Ken Murphy, National President of the Moon Society

◇◇◇

"Your book was a pleasure all the way. The review took me about 20 minutes to write." Dr. David Fischer, Vice President of the National Space Society Phoenix Chapter

◇◇◇

ISBN 978-1-938586-42-2
Revised Edition May 20, 2014
All Rights Reserved
Copyright © 2014 by Charles Lee Lesher

http://www.charleslesher.com
Published, May 1, 2012

eBook Edition June 4, 2014
ISBN 978-1-938586-71-2
Printed in the United States of America
Writers Cramp Publishing

http://www.writerscramp.us/
editor@writerscramp.us

◇◇◇

Out of the Cradle

Charles Lee Lesher

Space Based Solar Power

Abundant cheap electricity is a key element in getting and maintaining high human living standards around the globe. Stated another way, electricity is the foundation of modern technology. Without it, we go back to sailing ships and the horse. Anyone who thinks for a moment that we could feed and clothe our 21st century population using 18th century technology is nuts. Everything in your world depends on electricity either directly or indirectly. The food you eat, the water you drink, the car you drive, are all possible because of electricity. This cannot be overstated. Electricity is civilization.

The world consumed 20.1 trillion kilowatt hours of energy in 2010, up over 3% from the year before. Of this, 13.9 trillion kw-hrs comes from hydrocarbons, 2.6 trillion from nuclear, and 3.1 trillion from hydroelectric. Earth-bound renewables hardly make a dent. The EIA predicts world net electrical consumption will increase to 25.5 trillion kw-hrs by 2020 and 35.2 trillion kw-hrs by 2035. Even if you don't fully understand the enormity of the numbers, just remember that we are using more energy today than we did yesterday, by a lot.

There is only one power source capable of meeting this demand and all future demands, the SUN, and the best place to harness its power is collecting sunlight in SPACE. This book lays out why we need to do this, how to do it, and who's doing it.

OUT OF THE CRADLE

THE LOW ROAD

THE HIGH ROAD

CONCLUSION AND SUMMATION

Forward

One of my earliest childhood memories is the opening from the children's TV show Big Blue Marble. The show capitalized on one of the most iconic images from the Moon program - that of the Earth floating like a blue marble in the vast blackness of space. All of the lands. All of the waters. All of the peoples. Everything we know, all traveling together on a tiny planet in the vast void of the cosmos. While the idea is more commonly accepted now, at the time it was a significant shift in mind-set from what had gone before, a shift that spawned generations of environmental activists and laid bare the lie that we could endlessly exploit our planet without consequence. But our future is not without hope.

Modern civilization requires resources, so if we don't want to keep tearing up our planet, where do we get them? There is another heavenly body within our grasp, the Moon. Space activists are sometimes accused of wanting to escape Earth and its troubles. Quite the contrary, most of us want to help make a better Earth. Opening the space frontier will allow us to exploit the energy and natural resources of the Moon and cislunar space in order to minimize, and even begin to remediate, the damage we do to our precious Blue Marble. This is especially true with energy.

*The great majority of our energy production comes from burning hydrocarbons or splitting atoms. Within the pages of this book are many instances detailing what this is doing to our delicate Blue Marble but in the final analysis, it also presents a compelling solution, **Space Based Solar Power**. Using **SBSP** we can stop burning oil, coal, and natural gas, and still provide the benefits of modern civilization to the burgeoning billions of citizens worldwide.*

This is not science fiction, nor has it been for a while. There is no Law of Nature preventing us from developing space to better our Earth. But it will only happen by our initiative, our invention, our industry, and our investment. The first steps have already been taken. Among other things, solar powered satellites form an integral part of a global communications system that is unparalleled in human history. The next step is to harvest that same energy to power our civilization. A healthier planet is the greatest gift we can leave our children. Organizations like The Moon Society, National Space Society, and Space Frontier Foundation all work towards doing just that. Join us, or create your own local chapter. We can show you how you can help.

Welcome to the Future!

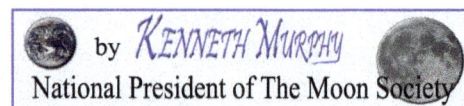

by *Kenneth Murphy*
National President of The Moon Society

The Low Road

Have a seat and let me tell you about an idea that will change the world. How's that for a grandiose opening statement? However, before you dismiss it and go on with your life, take a moment to understand what I'm talking about.

Who am I to think that I have any answers let alone one that will change the world? You're right. I'm just one of seven billion humans living on a tiny little planet in the corner of an obscure galaxy. I don't pretend to have all the answers, just one. Are you ready? Here it is… I know where there's an energy source capable of replacing coal, oil, gas, and nuclear in their entirety. Clean, safe and virtually unlimited, it is the ultimate renewable source.

The organization of this book is simple. In the first half, **The Low Road** paints a grim picture of what we are doing to our planet. Many smart people have already written about it, so let me apologize for bringing it up again. If it is any consolation, I don't dwell on any one thing for long, summarizing in a paragraph issues that require their own personal research library and years of study to truly understand. I have also tried to minimize the mathematics involved but alas, the numbers are such an integral part of a complex situation, I was not 100% successful. You will need your thinking cap and perhaps a barf bag for those with a delicate stomach.

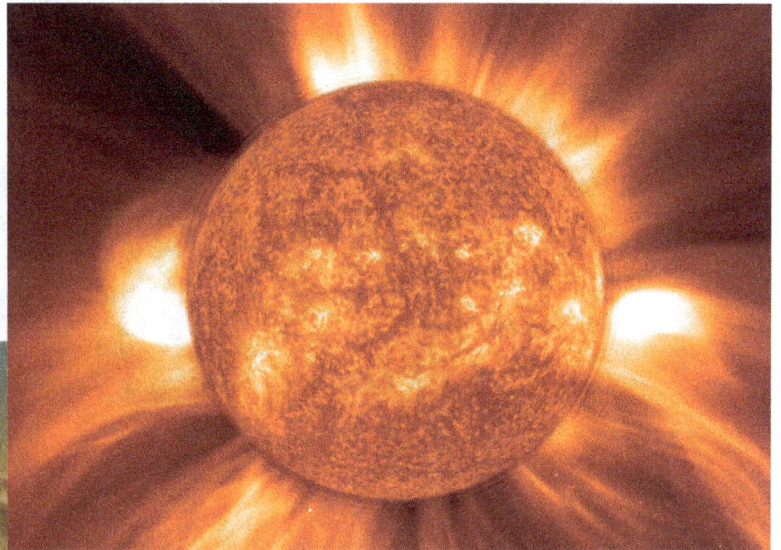

The High Road is where the fun begins. For those who have never heard of **Space Based Solar Power**, let this be your introduction to the power source of the future. For those who already adhere to its promise, let this book give you the tools you need to describe what **SBSP** is and intelligently discuss how we get there. It is one-stop shopping for those who care about the world we leave to our grandchildren.

That's it. I have told you what I'm going to tell you and at the end of the book, I will tell you what I told you. If that doesn't make your head spin, then turn the page and let's get started.

The View from the Cradle

Let's begin by stepping back. From the perspective of an extraterrestrial super-race looking down on us from above, Earth is dominated by one species, Homo sapiens. Humanity occupies every corner of our world… but it hasn't always been so.

In the past, our planet presented us with seemingly endless variety of new frontiers. That's no longer the case. There are no more new worlds to discover, no more strange and exotic lands just over the horizon. The world isn't flat any more.

Today it is common knowledge that the Earth is a sphere with a radius of about 6,300 km. Pictures taken from space reveal our blue little world in all its splendor. Water covers seventy percent of its surface leaving about 150 million square kilometers (km^2) of dry land for us. Of that total, ice covers 11 million km^2 and deserts consume another 6 million km^2. There are seven continents containing fifty-four major rivers, six major mountain ranges and hundreds of minor ones. There are five oceans with over a hundred and fifty seas and other named waterways. Mankind is intimately familiar with them all. There are no more new frontiers left for us to conquer down here on Earth. For that, we must look up.

When we do look up, we soon realize that for all its size, the Earth is but a speck of dust in the unimaginable vastness of the universe. We experience night and day because the Earth spins like a top. Days turn into years because the Earth is in orbit about the Sun. We can see the Moon, some planets and a few thousand stars with our naked eye. Beginning with Galileo, humans began building devices to help us discern more of what exists beyond what we can see. We now have the Hubble and other powerful instruments that let us probe ever deeper into space and back in time. What has become evident is that our Sun is but one of billions of stars that exist in a galaxy we call the Milky Way and beyond our galaxy is a universe containing many other galaxies. As we focus our telescopes even deeper into space and time, we see more galaxies seemingly without end, sharing several important facts only recently discerned but not understood.

Fact One: About 14 billion years ago[1], at a single point in space, matter/energy in a state of near-infinite density, near-infinite pressure and near-infinite temperature, surged into existence bringing its own spacetime with it.[2] Where before there was nothing, now there was something and that something was our universe. We humans gaze into the night sky and wonder at the magnificence of what happened next.

Area of a Sphere

$$A = 4\pi r^2$$

1 7/19/2010: NASA, Universe 101
2 7/19/2010: NASA, Astrophysics
http://www.nasa.gov/home/index.html

Age of the Universe

Big Bang

14 13 12 11 10 9 8 7 6 5 4 3 2 1

Billions of Years

Age of the Earth

4 3 2 1 500

Age of Multicellular Life on Earth

400 300 200 100

Millions of Years

Age of Man

Civilization

200 100 Now

Thousands of Years

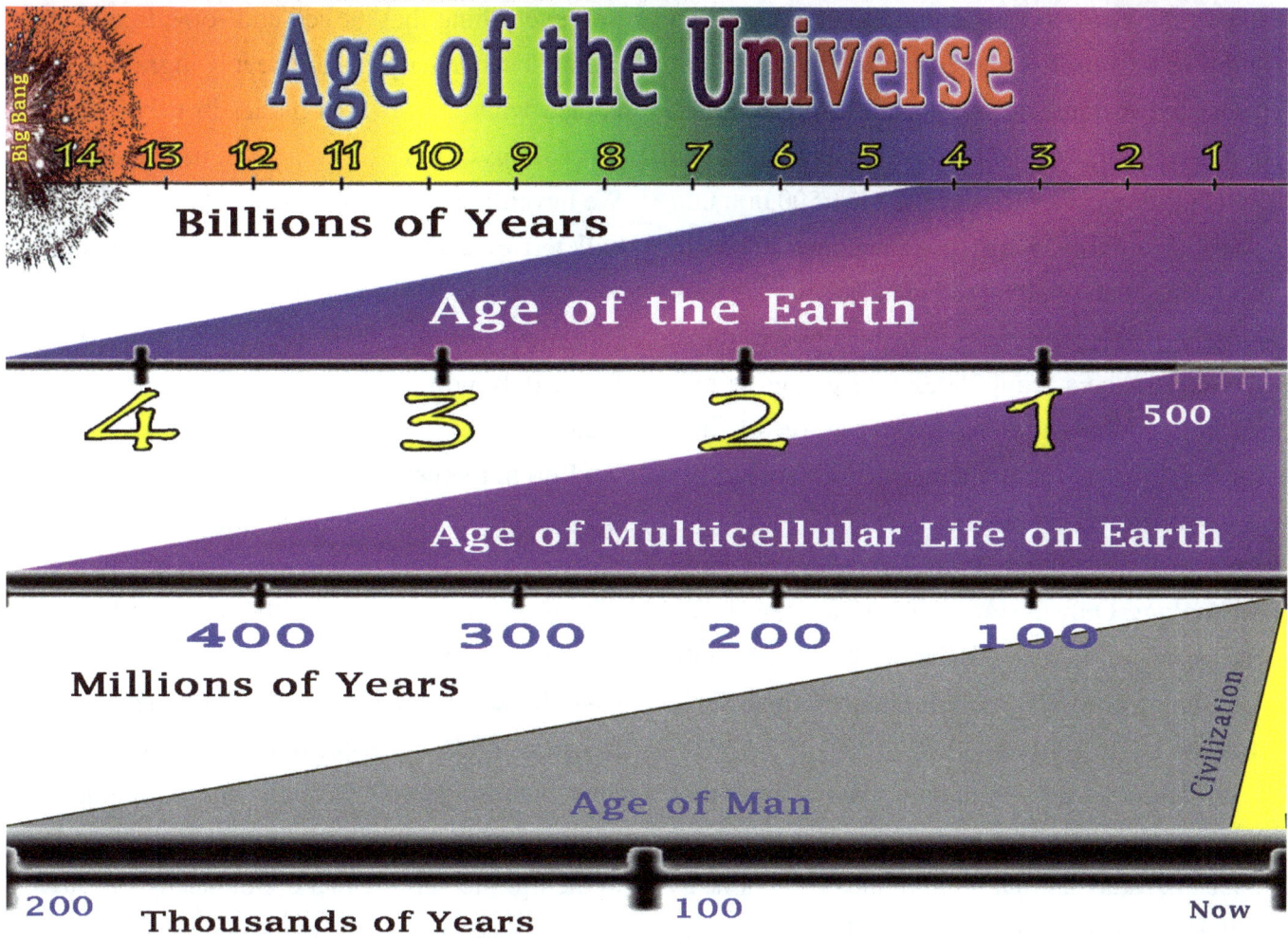

Fact Two: All galaxies, with the exception of those in our local group, are moving away from our galaxy like particles in a great explosion. At the time of this writing, humans could see about 13 billion years into our past with no measurable change in the density of the galaxies they find there, all moving away from us.

Fact Three: Our universe is not only expanding, but the rate of expansion is increasing.[3] Something is forcing the galaxies apart. We don't yet know what it is.

Fact Four: At the heart of every galaxy, there exists a supermassive black hole[4] millions of times the mass of our Sun.

Fact Five: The mass of the supermassive black hole is simply not massive enough to account for the high rotational speed of the stars in its galaxy.[5] In other words, galaxies rotate much faster than they should when applying just gravitational physics. Something else is going on. We don't yet know what it is.

What is evident is that we are just beginning to discern our place in the cosmos. Most of what we do know has been added to humanity's collective knowledge only in the last few centuries and we continue to learn more every day in an exponential curve.

We are a young species, not yet out of the cradle. Only recently have we determined the

---

3   10/4/2011: Nobel Prize for discovery of accelerating universe, Symmetry Breaking

4   9/15/2011: Small distant galaxies host supermassive black holes, University of California Santa Cruz

5   1/5/2011: Galaxy rotation curves, Tuorla Observatory, University of Turku

age of the Earth to be about 4,500,000,000 years old. For most of that incredible span of time, life consisted of single-cell organisms. More complex multi-cellular life didn't develop until about a half billion years ago. For the next 500,000,000 years, life evolved into a tremendous diversity of forms that finally produced us. If you look closely at the chart, Homo sapiens finally make an appearance a scant 200,000 years ago, a tiny sliver of time right at the end of the Age of Life, too brief for anything but the smallest scale.

Even for most of those 200,000 years, our ancestors were hunter/gatherers with very little technology. Only the last ten thousand years bears the mark of human civilization. Considerably less than a blink of an eye no matter which scale you use.

What separates humanity from every other species that has ever existed on planet Earth? Speech? Many animals communicate but none as efficiently as man. *Intelligence?* There are other creatures with large brains encompassing many complex adaptations but none with the abilities of ours. *Opposing thumbs? Walking upright?* Other species have these traits and do not build airplanes.

Perhaps it was our social skills that set us apart. After all, humans are slow, bad climbers, and having no claws or fangs, they can't fight back. All things considered, our ancestors must have made easy and tasty meals for predators. Instead, our ancestors learned the most important thing for anyone to know even today; individuals are easy prey but banding together made them all stronger. The whole is greater than the sum of the parts. This realization alone turned them into hunters. Suddenly there were packs of humans running around poking fire-sharpened sticks into everything that moved and some things that didn't. This underlying and deeply ingrained trait is also what drove us to gather in tribes, cities, and nations or as I like to think of it, Civilization. We haven't changed much since those times. We still terrorize the land whenever and wherever we want. However, some of us seem to have forgotten that together we stand, divided we fall… only now the game is played on a global scale.

I do not believe there's one single reason for the success of Homo sapiens. Rather, it is all of these characteristics converging and manifesting within us something that truly separates our species from all the rest, our ability to learn and pass this knowledge on to the next generation. Because of this, our ancestors began manipulating their environment through the imaginative application of technology, giving them an edge in the fight to survive that continues to this day.

Our technological evolution has defined who we are every bit as much as our biological evolution. It allowed us to adapt to ever-changing environments and situations that would have killed us otherwise. Ice ages and volcanoes, floods and droughts, locusts and epidemics, even the vacuum of space, have all been conquered with the aid of technology.

Reading and writing hold a very special place in the long list of human technology. Somewhere in the unreachable depths of time, there was a person who was the first to scratch lines in the dirt and give meaning to them beyond the physical. *I Salute You!* From such humble beginnings, great things grew. Without writing, there can be no mathematics. Without mathematics, there can be no science. Without science, we do not have civilization. Until writing, the wisdom of man passed from one generation to the next by word of mouth, a notoriously unreliable way to communicate.

I can understand the almost magical effect writing had on prehistoric people. Written words can paint pictures in a person's mind and were powerful messages. Without writing, religion would not exist. Writing gave the early church its power. Only priests knew how to read and write. Religious leaders enacted laws with severe penalties forbidding commoners to read the Holy Words or possess a Bible. They reserved that privilege for themselves. I don't believe it is a coincidence that the biblical age of the Earth and the invention of writing occur about the same time. However, for all its power, religion could not stop the accumulation of knowledge once started. Books became synonymous with learning. Writing necessitated the invention of education and the quest for truth began in earnest. Over the next nine thousand years, alphanumerics evolved from crude scratches in clay to the versatile symbols you are reading right now. Aren't we a clever species!

All my life I have enjoyed going to parks, stretching out on the grass and looking up at the trees. Some mighty fine oak trees grow on the University of Wisconsin campus in Madison, Colorado has elms, California has Joshua trees, and here in Arizona we have Palo Verde and Mesquite. When I look at a tree, or any living thing, I feel a kinship that extends back in time far beyond my meager ability to imagine. In the beginning, life consisted of a single species of one-celled creatures emerging from the chemical froth of a very young Earth, perhaps in a tidal pool stirred by the Moon and heated by the Sun. From this humble start, life evolved over a length of time impossible to grasp in its entirety, branching out and diversifying into all living creatures, past and present.

⬦⬦⬦⬦⬦⬦⬦⬦⬦⬦⬦⬦⬦⬦⬦⬦⬦⬦⬦⬦⬦⬦⬦⬦⬦⬦⬦⬦⬦⬦⬦⬦⬦

"It is not the strongest of the species that survives, nor the most intelligent that survives. It is the one that is the most adaptable to change."
Charles Darwin

⬦⬦⬦⬦⬦⬦⬦⬦⬦⬦⬦⬦⬦⬦⬦⬦⬦⬦⬦⬦⬦⬦⬦⬦⬦⬦⬦⬦⬦⬦⬦⬦⬦

The tree and I share a spot on the leading edge of this vast wave of life. We are both the current manifestations of a recently recognized process we humans call evolution. However, it does not end with us. We are not a final product. Instead, the tree and I are part of an awe-inspiring continuum through space and time that marches on with each successive generation. All two-hundred-thousand years of human existence is but a freeze-frame in this immensity, a flicker in time when a species gained the wit to ask questions, not about gods and the afterlife, but about our true place in the universe. This is a wondrous and beautiful story, one that everyone can take pride in. We are special, but so are all other living things that share the Earth with us, even the trees. We have a responsibility to care for the world and not abuse it.

In our brief but tumultuous history, humanity has risen from ignorance to prominence in an incredibly short time. No other species has ever dominated the planet as thoroughly as we do now. However, the very characteristics that made us successful now threaten everything we have built.

Go Forth and Multiply

The single overwhelming fact driving most of our problems is human population pressure. Sheer numbers is stressing our planet to the breaking point.

The graphic on the previous page clearly shows the exponential nature of human population growth. What it doesn't show is how close we are to overwhelming the Earth's ecosystem. According to the United Nations report, *Determinants and Consequences of Population Trends*, the number of Homo *sapiens* grew very slowly or not at all for most of human history. What little is known about this distant past comes from intensive study of bones and artifacts. Without our ancestor's census data, we don't know exactly how many of us there might have been, but by 10,000 BCE, best guess is under 5 million people, worldwide.[1]

increased to about 500 million. It wasn't until 1800 that we cracked the first billion. In 1925, it topped 2 billion but by 2000, the planet's population had exploded, reaching over 6 billion. As of the 2010 census, America alone contained over 300 million citizens. The world surpassed 7,000,000,000 sometime around midnight on Halloween 2011. (1000 million = 1 billion)

At the risk of going mathematical on you, let's take a closer look at what that means. A decent approximation of the annual Growth Rate over some period is simply the ratio of the ending population minus the starting population divided by the starting population. The graph[2] below shows the world's Growth Rate from 1964 to 2012 plot with the population. From the recent high of 2.11%, the world's current Growth Rate is at 1.12% per year.

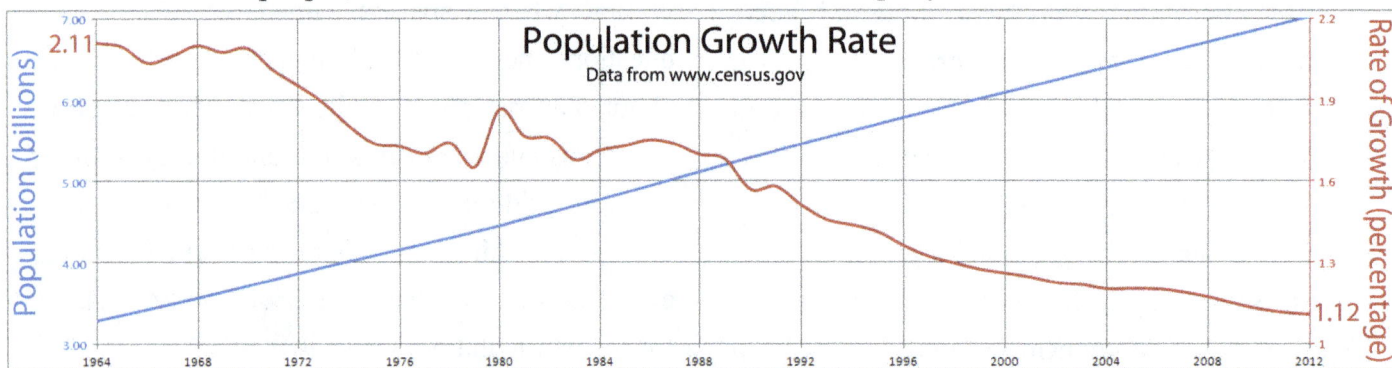

Population Growth Rate
Data from www.census.gov

Between the agricultural revolution and the Roman Empire, population Growth Rate was less than a tenth of a percent per year. Sometime in the 1st century, we reached 300 million. Then plagues began sweeping across the land. The largest was the '*Black Death*' in the 14th century that wiped out at least 75 million people. Despite everything, by 1650 the world's population had

If you know our Growth Rate, then even a fifth grader can calculate how long it will take the Earth's human population to double in size using the Rule of Seventy.[3] This simple formula approximates a more complicated and exact doubling function. The Rule of Seventy is simply seventy divided by the growth rate percent. Enough math for you? What's important is that

1 3-1-2007: Fact or Fiction? Living People Outnumber the Dead, Ciara Curtin

2 6/1/2011: World Population Growth Rates, U.S. Census Bureau

3 5/17/2002: Exponential Growth and The Rule of 70

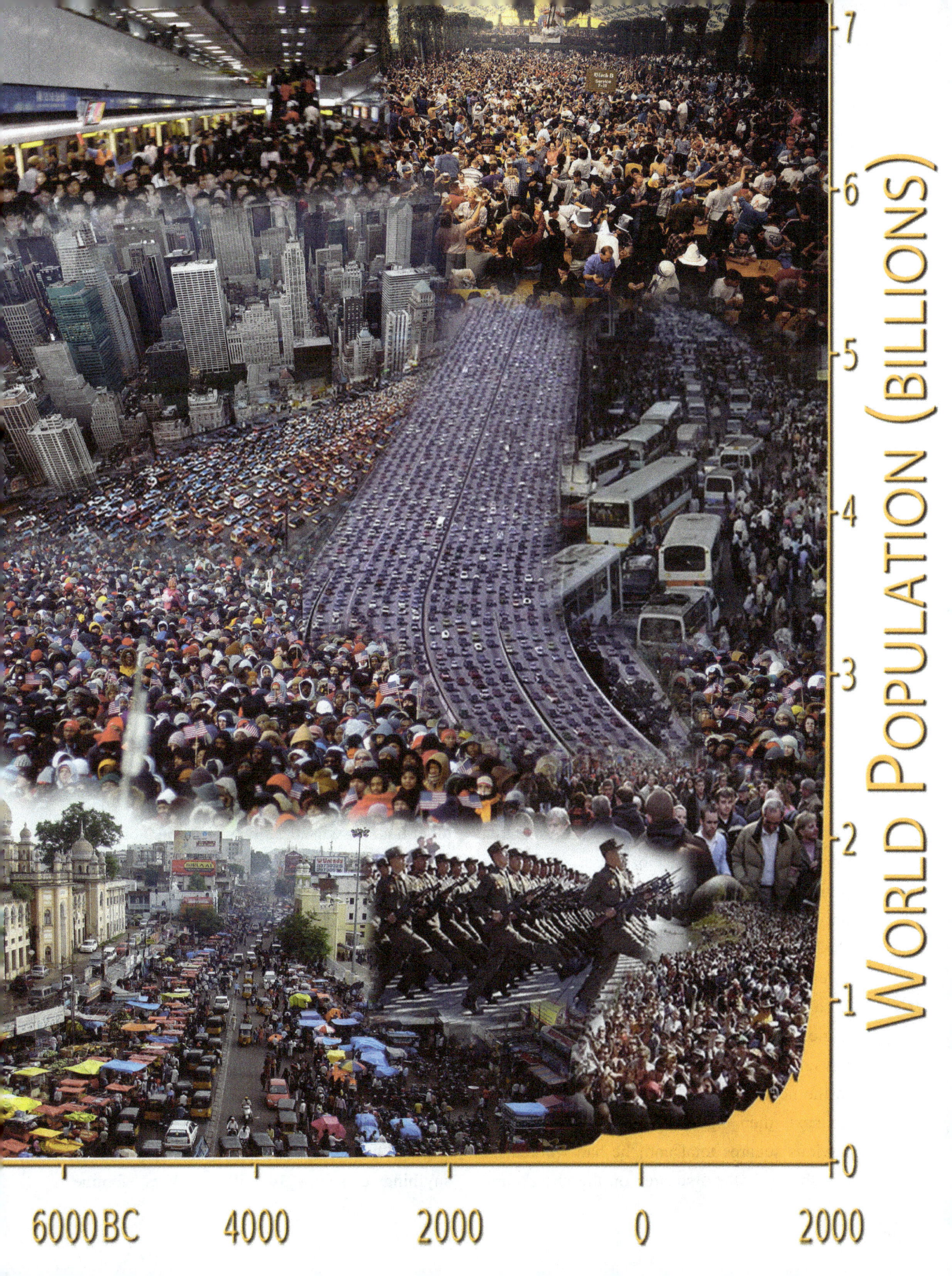

WORLD POPULATION (BILLIONS)

7

6

5

4

3

2

1

0

6000 BC   4000   2000   0   2000

this simple equation gives the time it takes for a given population to double in size.

Plug in 70, divide by 1.12%, and you will get 62.5 years. That's right. At our current Growth Rate, the world's population will double in a single lifetime, from 7 to 14 billion people. I grant you, this is the high water mark. Many things will happen to lower the Growth Rate but the point is, there are already a lot of people sharing this world with you with a lot more right around the corner. Keep this in mind as we take a good look at what seven billion people are doing to our beautiful Mother Earth right now. In the years to come, the pressure to support so many will become enormous. Sustainability must always be paramount in deciding what path to take to ensure a future for our children.

To illustrate the power of doubling, let me tell you a little story. Once upon a time, there was a king who was greatly pleased with something completely new, the game of chess. To show his gratitude, he offered the man who invented this marvelous new game a reward of his choosing. The king expected him to ask for land or gold or jewels but instead the inventor of chess thought for a moment and asked the ruler to place a single grain of wheat on the first square of his new chessboard. Double it and place two grains on the next, then double that to four grains on the next, then eight and so on, doubling the previous squares total until he had worked his way through all 64 squares on the chessboard.

The king granted him the wish until he realized just how much wheat was involved. The number of grains needed just for the last square alone is 9,223,372,036,854,780,000. That is about 400 times the 1990 worldwide harvest of wheat!

The point is that simple doubling can generate huge numbers very quickly. Let me repeat. If human population continues at the current Growth Rate, it will double within the next 60 years. That means the only thing standing between you and sharing the world with 14 billion of your closest friends is a single lifetime. Is that even possible? Can the Earth feed so many? For how long? Where will all the electricity to provide for them all come from? Coal? Nuclear? Gas? Solar? Wind?

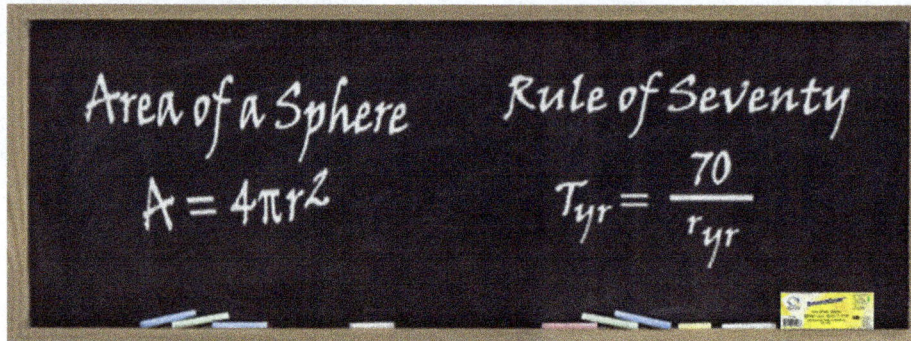

Within the United Nations is the Department of Economic and Social Affairs, Population Division, charged with predicting how many other humans will share the world with you in the near future. After extensive studies, they came up with the High and Low curves representing the range of likely outcomes. Referring to the graph, the Best scenario is the UN's calculated best guess as to what will actually happen. Plotting the 2010 census data on the 2004 graph, it becomes apparent that the actual population numbers are closely following the Best curve, just as predicted.

From the beginning, the UN Long-Range Population Projections never seriously predicted anything close to 14 billion. Wars, famine

Area of a Sphere

$$A = 4\pi r^2$$

Rule of Seventy

$$T_{yr} = \frac{70}{r_{yr}}$$

and other disasters will reduce the Population Growth Rate substantially long before we reach that lofty figure. The original projection along the Best variant was 9.4 billion but recent studies have the UN revising their estimates downward even more. The UN now predicts the world's population in 2050 to top out at 8.9 billion.[4]

subcontinent. The most important factor is HIV/AIDS, which is spreading much faster than previously anticipated, thus increasing the Death Rate.

Regardless, in 2011 the number of humans living on this one little planet exceeded seven billion. Within my granddaughter's lifetime, the

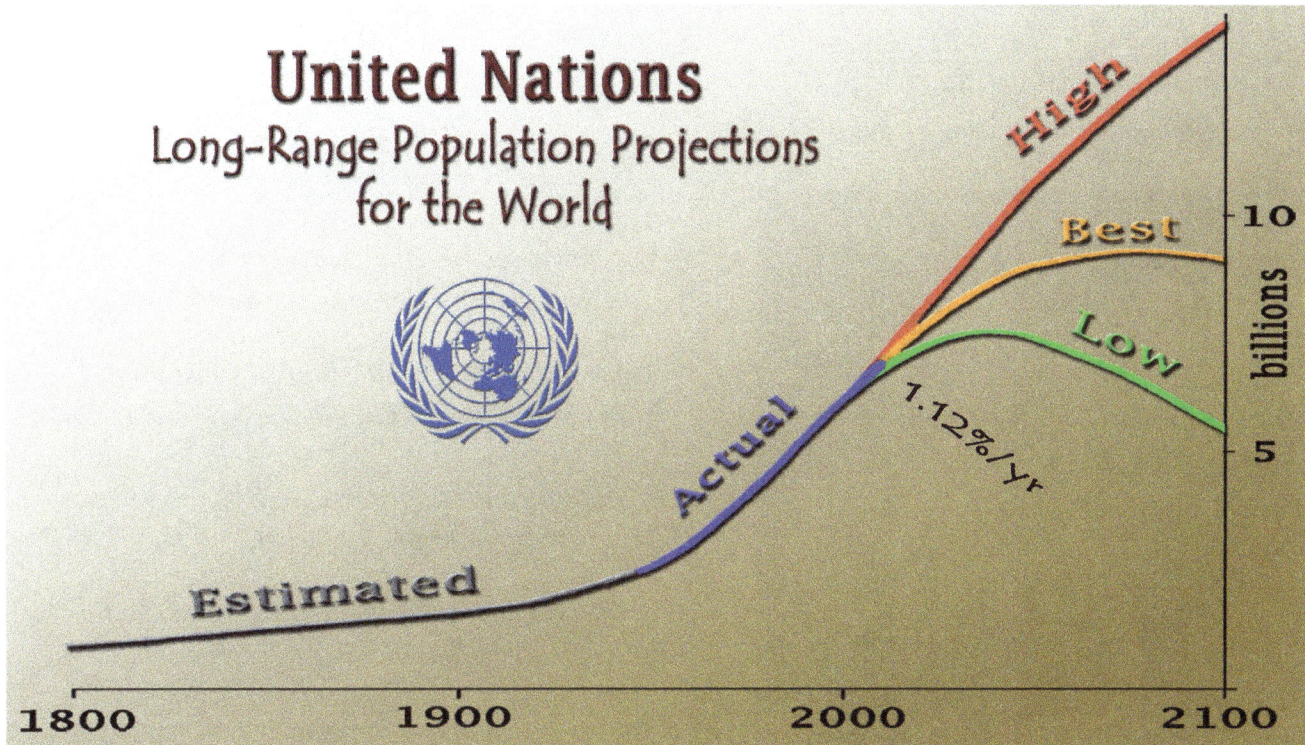

United Nations
Long-Range Population Projections for the World

High

Best

Low

Actual

1.12%/yr

Estimated

billions

10

5

1800   1900   2000   2100

The major reason for the lower figure is good news. Global fertility rates have declined more rapidly than expected as health care in impoverished nations has improved faster than anticipated. Simply put, as the number of babies that survives goes up, mothers are choosing to have smaller families thus, lowering the Birth Rate. That's what happens when civilization comes to a third world nation. Education, electricity, and women's rights are the leading factors playing a role here. However, about one-third of the reduction in Population Growth Rate is attributed to increasing mortality rates in sub-Saharan Africa and parts of the Indian

UN predicts another two billion people will join us. The population density in your neighborhood will vary but the UN report states the number of people per square kilometer of land will range from 504 persons per square kilometer in Micronesia to 3.6 in Australia/New Zealand. Some large countries in South-central Asia will become unusually dense by 2100. India is projected at 491 citizens per square kilometer, Pakistan 530, and Bangladesh a jaw dropping 1,997 people per square kilometer. The pressure to feed and house so many will stress our planets ecosystem to its breaking point and beyond. My friend and poet Harold Saferstein says it best:

---

4   World Population Prospects: The 2004 Revision, United Nations

Overpopulation

A Poem by Harold Saferstein

Scientists agree, and have quite clearly stated
That our planet's already over-populated

The uneducated masses add to these ills
They're not taught about using birth control pills

Religious leaders compound this deception
Discouraging all forms of contraception

For unwanted pregnancies, abortion's discouraged
And family planning is never encouraged

Some politicians I know think it's just great
To point to their kids, all seven or eight

Population control does not merit their concern
Millions already starving, ah, when will they learn?

Our air and our water are defiled with pollution
Yet our leaders don't bother to seek a solution

We are seven billion people and increasing too fast
Earth's resources we know cannot possibly last

But the priests and the mullah's are preaching it still
These very large families are fulfilling God's will

China got it right when they passed a decree
Limiting only one child to each family

Just how long can this planet survive, I wonder?
Doesn't matter to me, I'll soon be six feet under

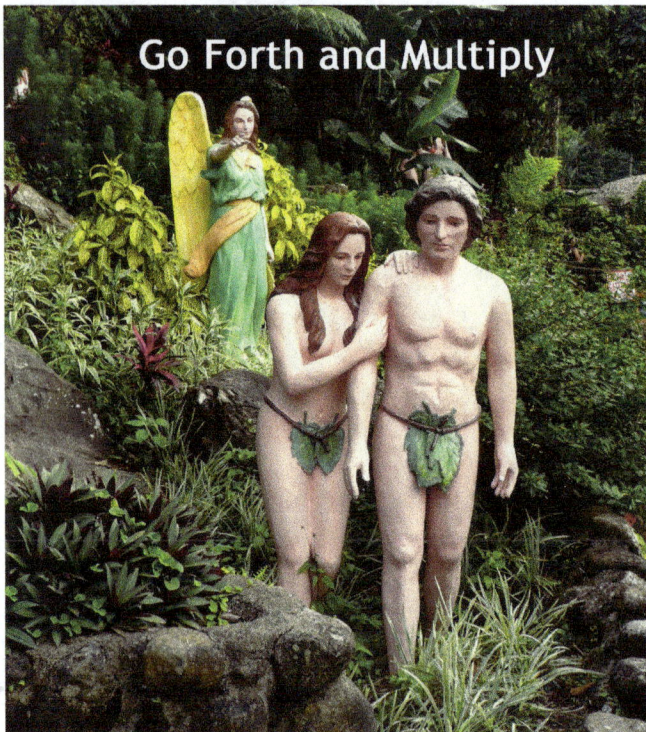

Go Forth and Multiply

Enough Already!

7 BILLION AND COUNTING

Masters of Genetics

The rate of scientific discoveries has been increasing exponentially for centuries and we are on the leading edge of that vast wave. New technology comes at us so fast that most people have become numb to it. New electronics? *Ho hum*. A revolutionary physics theory, another extraterrestrial planet, or uncovering the bones of a hitherto unknown ancient ancestor, nothing fazes us anymore. We accept world-changing discoveries matter-of-factly, that is, if we notice at all.

However, something special is happening in the field of genetics. Historically, genetics is the study of heredity, but over the last half century, it has become synonymous with understanding and controlling DNA. As such, genetics is a subject that holds the almost magical promise of human health and longevity. Even though we have only barely started down this path, genetically modified plants and animals help feed billions worldwide,[1] doctors routinely screen for genetic abnormalities in newborn babies,[2] and criminologists use DNA to identify suspects.[3] However, where is all this science taking us? Is anyone driving the bus?

In order to understand where we are going, we must first look back at where we have been. In 1665 using a primitive microscope, Robert Hooke described honeycomb-shaped structures he found on a piece of cork. He called them cells. In the 1830's, Robert Brown used a vastly improved microscope to observe a central sphere inside cells. He called this structure a nucleus. Soon thereafter, Theodor Schwann and Mathias Schleiden concluded that the nucleus plays a central role in the reproduction and growth of cells, thereby establishing the basis of modern cell theory. Almost two centuries after Hooke and some thirty years after Brown, in 1859 Charles Darwin published *The Origin of Species* in which he presented the theory that those members of a population who are better adapted to their environment survive to pass on their

Charles Robert Darwin (1809 - 1882)

traits. However, Darwin could only describe the macroscopic result of the evolutionary process, not how it worked on a molecular level.

Knowledge accumulated steadily over the next half century so that by the early 1900s, a hereditary hypothesis was taking shape and gaining acceptance within the biological

---

1   3/23/05: Biotech corn mistakenly sold to farmers. Associated Press
2   Newborn Genetic Screening
3   10/2/2005: Bill would expand DNA database. USA Today

community. Researchers identified extremely long threads of nucleic acids and proteins residing in the nuclei of every living cell. They called them chromosomes. Somehow, this enormously complex arrangement directs the growth of the individual and the evolution of a species.

It took until 1953 for James Watson and Francis Crick to develop the three-dimensional model of a DNA strand, the now famous, double helix. Let's dwell on this for a moment. Two years before I was born, scientists identified for the first time what the inner core of life itself looked like. Only then, could they even ask the right questions, let alone ferret out the answers. What part of DNA controls growth? How does it work? Can it make a mistake? What role does DNA play in disease? In aging?

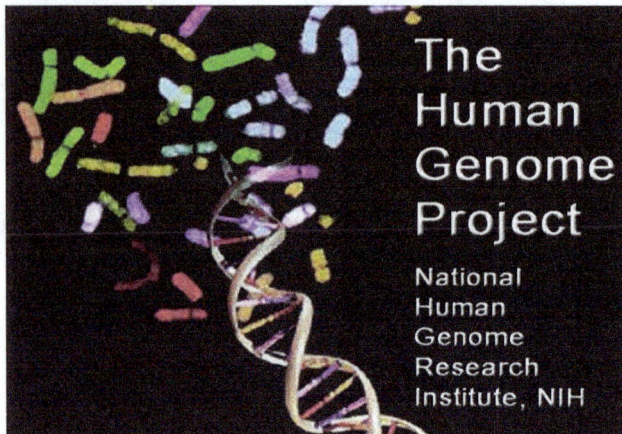

The Human Genome Project

National Human Genome Research Institute, NIH

In the first major effort to understand our own bodies, the Human Genome Project (HGP) was launched in 1990. It was an international $3 billion effort to sequence all three-billion base pairs and map the genetic markers on all 23 chromosomes in a single human genome. Thirteen years later, on the 50th anniversary of the discovery of DNA's double helix structure by Crick and Watson, HGP published the first complete human genetic sequence in the journal Nature (24 April 2003), more than two years

ahead of schedule. Since then, technology has improved. Today, researchers survey a half million base pairs at a time making it possible to map an entire genome in hours for only a few dollars. This amazing progress is opening up an entire new line of investigation.

international BARCODE OF LIFE

The International Barcode of Life (iBOL[4]) is the largest biodiversity genomics initiative ever undertaken. It uses a very short genetic sequence from a standard part of the genome to positively identify a species. The comparative data is kept in a global reference library that will eventually contain DNA from every species on Earth and even some that are extinct. People and organizations are already using this grand directory for a variety of purposes ranging from identifying processed meat shipped from half way around the world, to behavior of birds around airports.[5] As more and more genomes within a species accumulate in this enormous repository, the better we understand the subtleties within that species.[6] Human DNA is but one among the multitude, special to us, yet realistically simply another thread in the fabric of life that exists on our little blue planet.

However, this is not the only such project. There are many collecting and studying DNA. The HapMap Project is an international effort aimed at collecting human DNA from populations all over the world.[7] The National Geographic Society sponsored genomic studies to shed light on the expansion and evolution of ancient humans.[8] Using comparative analysis, researchers are just beginning to mine this

4 International Barcode of Life: http://www.barcodeoflife.org/
5 9/15/07: Government taps into DNA barcodes for 1.8M species. AP
6 9/6/06: Genetic family tree of all life is beginning to bear fruit. National Geographic News,
7 International HapMap Project
8 A Landmark Study of the Human Journey

immense ocean of data, striving to identify the subtle differences that define the characteristics of living organisms. Make no mistake, complete understanding is coming, it is only a matter of time.

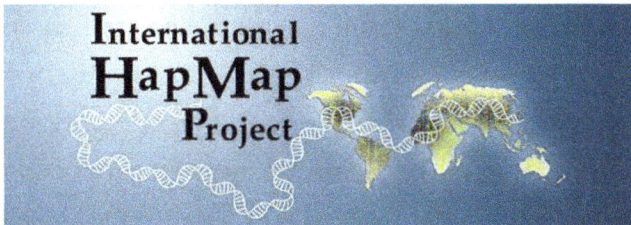

What does this mean to you and me? Researchers have already identified the gene variants causing macular degeneration, breast cancer, prostate cancer, colon cancer, and inflammatory bowel disease.[9] Of particular note, they are studying plants that can mend their own genetic mistakes.[10] Scientists routinely clone genes in the course of their studies,[11] and have even extracted genetic material from a 38,000-year-old Neanderthal bone fragment.[12] The amount of information concerning genetics is already astounding and growing every day. As scientists learn, family doctors will expand the number of genetic therapies to include virtually every ailment, curing everything from being a blind overweight moron with a bad memory to cancer, diabetes, and AIDS. In the near future, you will have a genetic therapy designed just for you, based on your personal genome.[13] Doctors will soon have almost total control over the biology of living organisms, human or otherwise, able to treat their patients at the molecular level. Scientists have already stimulated bone marrow stem cells to grow into a complete, functioning heart valve for children in a process called Tissue

Engineering.[14]

Are we nearing the end of this journey? Will we wake up tomorrow and have all the answers to human health and longevity? Probably not. The more we discover, the more we realize just how complicated life truly is. Your DNA is unique between you and the next guy but not within your own body. Every cell contains your DNA. How does your body know when to grow a skin cell instead of a blood cell? What changes between a heart muscle and a liver cell? The answer is the epigenome. Where our genome provides the blueprint, the epigenome provides the guidance. Epigenetics is the biological control mechanism that tells DNA what to do and when to do it. It turns vast sections of base pairs on and off making it possible to grow a liver, heart, or brain cell. It dictates when and where they are to grow. Epigenetics is also the reason identical twins can be so different. One twin can be normal in every definition of the word while the other is afflicted with autism, yet, they have exactly the same DNA. We are only beginning to unravel epigenetics but understanding will come. After all, why treat a symptom when you can fix the problem at its source?

We have come to realize, biologically speaking, that genes control everything. Our growth, our metabolic rate, our mental processes, our health and thus, our happiness… and humanity is on the verge of understanding how it works, all of it, right down to the miniscule differences between you and me.[15] The rate at which these and other breakthroughs are reported is increasing exponentially, until today, we read about them almost daily. Here are but a few of the headlines: *Scientists use gene*

9 8/3/07: Human genome project is beginning to bear fruit, McClatchy News
10 2006 Plants can mend genetic mistakes. Washington Post
11 11/13/05: Scientists clone disease-causing gene. Tribune
12 11/16/06: Neanderthal bone could yield gene link to humans. Associated Press
13 10/18/06: Personalized medicine promises tailor-made diagnosis, treatments. National Geographic News
14 9/15/07: Scientists near heart valve breakthrough. Newsday
15 9/4/07: DNA suggests we're less alike than scientists thought. AP

therapy to restore deaf mammals hearing.[16] *Key gene believed root of evolution.*[17] *Gene studied for possible role in adult diabetes cases.*[18] *HIV gene tests encouraging.*[19] *Parkinson's gene therapy shows promise.*[20] *Genes research looks for clues to good memory.*[21] *Fat gene found but no clue how it works.*[22] *Gene tied to form of Alzheimer's.*[23] *Disabling of genes aids in cancer fight.*[24] *Study finds new genetic risk factors for Type 2 diabetes.*[25] *Researchers discover itching gene.*[26] *Scientists identify MS-causing genes.*[27] *Glaucoma-causing gene mutations discovered.*[28]

As more and more of this marvelous biological process falls under human control, what will we do with its power? Will we design our children? Sorry but that is already underway. For years, prospective mothers have had the ability to quickly and cheaply gender test their embryos.[29] However, what began as gender selection has quickly grown to include many of the genetic markers discussed earlier. Within the first few weeks of a pregnancy, parents can use this information to decide if they want to have an abortion and start over. Not surprisingly, some parents with deafness, dwarfism, or gigantism want similar kids.[30] They use this technology to screen for their particular genetic variant, establishing their own definition of normal.

Perhaps some among us will use this power to design the perfect soldier, someone incapable of questioning orders and willing to die on command. As a veteran and an observer of modern politics, I have no problem imagining Army generals or Washington politicians using genetic engineering to achieve their goals, all the while feeling justified in doing so. I hope this is simply speculation on my part, but it worries me.

However, there is a dark side to genetics. Sure, we would all like to live long lives, but what happens if we actually do stay healthy and vibrant well over 100 years. This is not as farfetched as it once was. Scientists have already discovered the genes that control aging in mice.[31] It is a very short step to finding the same markers in humans.[32] How will society change if genetics can extend human life indefinitely? How long can a human being live if their body never grows old? Perhaps someday we will control the aging process so well that we can back it up, have our doctor dial in what age we want to be. Imagine for a moment what it would be like to always be twenty-five, to live in a world where everyone else is twenty five. The birth rate must plummet to the point of stopping altogether. Otherwise, the humans would very quickly overwhelm the Earth's ability to feed them.

As a humanist, I respect knowledge yet am wary of those who would abuse it. No doubt, the genetic genie is out of the bottle and cannot be returned. Whether we like it or not, genetic

16 2/14/05: Scientists use gene therapy to restore deaf mammals hearing. LA Times

17 8/17/06: Key gene believed root of evolution. Associated Press

18 1/16/06: Gene studied for possible role in adult diabetes cases. Washington Post

19 11/7/06: HIV gene tests encouraging. Associated Press

20 4/17/07: Parkinson's gene therapy shows promise. Associated Press

21 3/17/07: Genes research looks for clues to good memory. McClatchy News

22 4/13/07: Fat gene found but no clue how it works. Associated Press

23 1/15/07: Gene tied to form of Alzheimer's. Associated Press

24 4/12/07: Disabling of genes aids in cancer fight. Dallas Good Morning News

25 4/27/07: Study finds new genetic risk factors for Type 2 diabetes. AP

26 7/27/07: Researchers discover itching gene. St Louis Post-Dispatch

27 7/30/07: Scientists identify MS-causing genes. South Florida Sun-Sentinel

28 Glaucoma-causing gene mutations discovered. LA Times, 8/12/07

29 Designing your own baby. Boston Globe, 8/8/05

30 Making designer babies – with genetic defects. Associated Press, 12/22/06

31 Scientists target souped-up gene to slow aging in mice. Seattle Times, 5/9/05

32 'It's the Holy Grail of aging research'. Associated Press, 11/2/06

science is quickly approaching the point when human life can be extended indefinitely. Will we grant this boon to everyone? Or will it only be for Bill Gates, Rupert Murdock, and others who can afford immortality? On one hand, I believe that health and a long life must not become the purview of the very rich and powerful. On the other hand, as the Death Rate plummets, the world's population will skyrocket unless birth rates drop just as drastically. Even if restricting birth rates worldwide were possible, a low rate would disturb the timeless cycle of life and death, causing generational change to grind to a stop. It is enough to make your head spin.

"I don't think the human race will survive the next thousand years, unless we spread into space."

Stephen Hawking

There are currently over seven billion people on the Earth, increasing by a billion more every dozen years or so. Simply maintaining this population explosion will quickly overwhelm the Earth's natural resources. Radical advances in medical science exert a positive influence on human health and longevity thereby exacerbating the situation. Yet, we cannot help but strive for the infamous Fountain of Youth without regard for the consequence of attaining it. Careful what you wish for, you may get it.

Staying confined to one little planet is a path leading to a grim future, a future where humanity overloads the Earth with sheer numbers, fighting for the last few scraps of food, water, and land. Having lots of money would help you survive this chaos and afford life-extending medical technology, but by necessity, the vast majority

of humanity will need to be excluded from these benefits. After all, we can't have everyone living forever. There simply isn't room. Can you doubt that money will determine who lives and who dies?

"(Space programs are) a force operating on educational pipelines that stimulate the formation of scientists, technologists, engineers and mathematicians... They're the ones that make tomorrow come. The foundations of economies... issue forth from investments we make in science and technology."

Neil deGrasse Tyson

There is only one viable solution providing all of humanity the freedom to have kids **and** enjoy long lives; expand into space in a big way! Bring the Moon and cislunar space, the region around the Earth and Moon, into the human sphere of influence. Then, after we learn the lessons that will teach us, go on to colonize the solar system beyond. There is no limit to what we can do after that. Seven billion people may just be a good start! Both Stephen Hawking and Neil deGrasse Tyson believe space colonization can save humanity. I'm simply pointing out another very good reason why they are right.

As luck would have it, Genetic Engineering is a very important component to any space colonization effort. The space environment is fraught with hazards that only a complete understanding and control of genetics can overcome. Zero gravity bone and muscle deterioration, radiation, and vast distances all become solvable with advanced genetics.

If you do manage to reach the fine age of 65, a woman living in America can expect to live

another 21 years compared to only 18 for men. Keep in mind that over a quarter of people aged 85 years and older suffers from dementia.

Extending and improving the quality of life is a good problem to have but nevertheless, it is a problem and we need reasonable people

Life Expectancy After Age 65

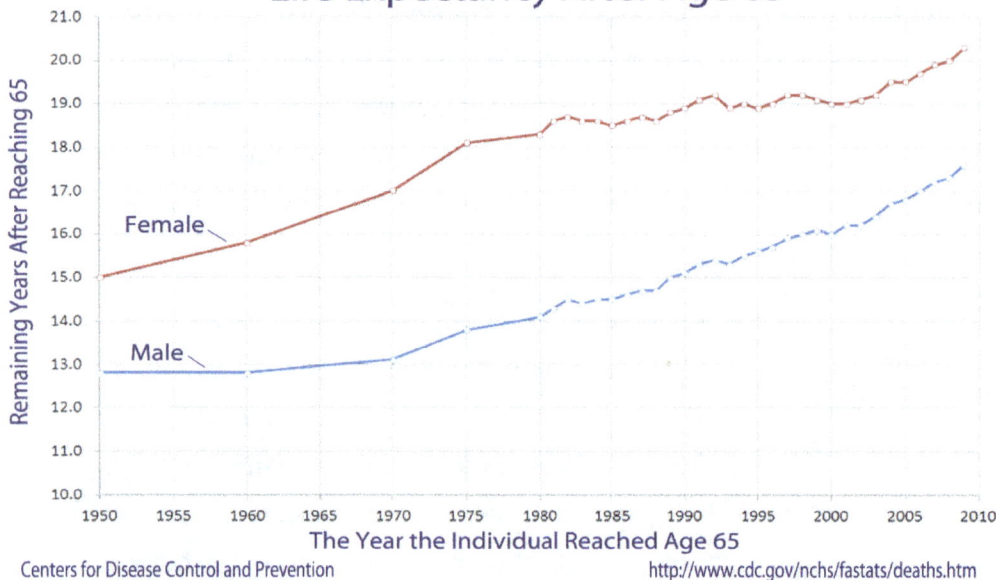

Centers for Disease Control and Prevention   http://www.cdc.gov/nchs/fastats/deaths.htm

discussing it openly and honestly. Outlawing stem cell research, or any research for that matter, is not the right approach. It simply shifts the problem away from America and pushes it outside our sphere of influence. What's forbidden in one country is encouraged in another. We are part of a global economy and research into genetics will take place somewhere. If we make laws against something here, it simply pops up somewhere else, but it does not go away, no matter how much our politicians think differently. We need sensible laws governing all aspects of this rapidly emerging technology, laws that protect the rights of everyone regardless of their station in life. But mostly, we need the collective will to colonize space and expand humanity's horizons. Only out there will we find the freedom to indulge in immortality.

Facts and Politics

On our way to Las Vegas recently, my wife and I noticed a dirty brown cloud hanging over one of the picturesque valleys far from any big city. With a shock, we realized it was smog, like what we see every day over Phoenix. It drove home the fact that atmospheric pollutants do not go away just because we no longer see them. They just change location, spreading out from their point of origin to stain the sky indiscriminately.

Farther along on our journey, we received another shock at the low water level of Lake Mead. From the vantage point of Hoover Dam, we could see the bathtub ring of mineral deposits painting the reservoir walls white, clearly marking the normal water level of just a few years ago. In the last six years, the lake has dropped 100 feet. If it drops 50 feet lower, the Southwest loses a vital water source and Hoover Dam stops generating electricity for California, Nevada, and Arizona, both massive economic disasters. The Bureau of Reclamation projects the lake to fall another 19 feet over the summer.

As a young man, I spent three years of a four-year U.S. Army enlistment stationed in Germany. While there, I skied the Zugspitze Glacier, part of the Garmisch-Partenkirchen recreation area. Today, the German people are struggling to save it.

Over the last two decades, the Zugspitze has lost half its mass[1] , going from 80 meters thick

---

1   5/1/2011: Alpine glacier covered in canvas to prevent summer melting

to less than 40. Every spring they spread huge tarps across the surface of the glacier trying in vain to save this ancient ice from melting away. However, they cannot stop the meltdown and the most optimistic among them estimate Germany's last glacier will be gone within another two decades. Closer to home, the 150 glaciers in Montana's Glacier National Park has shrunk to only 25.[2] The rest are gone and those that remain are melting. The fact is that the worlds' glaciers are disappearing at an unprecedented rate. After tens of thousands of years, why is this happening now? Can humanity be responsible for all or part of this? Our climate scientists say yes.

The relevant facts are that we pump 87 million barrels of oil out of the ground, mine 13 million tons of coal, and suck 200 billion cubic feet of natural gas out of Mother Earth *EVERY DAY*. We burn the majority of all these hydrocarbons, dumping enormous amounts of carbon and other pollutants into our atmosphere. In 2010, the global output of heat-trapping carbon dioxide increased at a record setting 6% with China, America and India the world's top producers. Economists say this is a sign that the recession is ending and energy demand is rising.[3]

That's not the only thing polluting our atmosphere. Growing worse every hot summer, forest fires spew even more contaminants into our air.[4] It has become an annual ritual to send forest firefighters into the hotspots. In 2011, Arizona and Texas both experienced record wildfire seasons.[5] The Amazon jungle is losing an average of three million acres of forests a year, an area the size of Mississippi, to slash and burn farming.[6]

In fact, 2011 was the worst year ever for billion-dollar weather disasters in America. We endured a truly biblical onslaught of tornados, floods, blizzards, drought, heat waves and wildfires. Three very dear friends who moved to Missouri a few years back were on the fifth floor of the six-story hospital in Joplin when the EF5[7] tornado hit the building and ripped through the town killing 160 people. My friends were lucky. They escaped with only superficial cuts and the indelible memories of sheer terror. The power of Mother Nature has a way of humbling the strongest among us.

Joplin was only one of six major outbreaks of tornados in 2011 killing a total of 540 people. In just four days in April, there were 343 tornados, the largest outbreak *ever*. They recorded 199 in one day, another record. The degree of devastation these tornados inflicted on America was extreme.[8]

Oklahoma set an American record for the hottest month *ever*. In my home state of Arizona, the ten-year drought worsened in 2011. Arizona gets only about seven inches of rain in an average year but in 2011 it plummeted to half that.[9] At the same time, the Ohio River Valley tripled its normal rainfall causing major flooding along the Mississippi River.

Climate scientists say 2011 is not just an anomalous year but a harbinger of things to come. What took the Earth millions of years to contain, we are releasing back into our environment essentially overnight, where, according to the Chairman of the Goldwater Institute in Phoenix, Arizona, all this CO_2 does nothing. I must conclude that the Chairman, and those that agree

2 4/8/2010: Two more glaciers disappear from Glacier National Park, AP, Mat Brown
3 11/4/2011: Global-warming gas sees biggest jump ever, Wire Service, AZ Rep
4 National Interagency Fire Center
5 1/19/2012: Wildfire Season 2011
6 Farming the Amazon, National Geographic
7 Enhanced Fujita Scale (0-5 with 5 being catastrophic)
8 Billion Dollar US Weather/Climate Disasters, NOAA
9 11/28/2011: Arizona drought worsens; 2011 rainfall totals low, Wire Service, Arizona Republic

with him, are not interested in reason, logic, or scientific facts, just in pursuing points for their particular political viewpoint without regard for the absurdity of their argument or the results if they are mistaken. After all, we will all be long dead before humanity suffers the full effects of global climate change. Just because it will not affect you is no excuse for ignoring the signs that something is terribly wrong.

Sometimes all it takes for a global-warming skeptic to change their mind is to honestly examine the data. Really, it happened. Richard Muller, a Cal Berkley professor of physics and long time global-warming skeptic, spent two years and a chunk of money from the Charles Koch Foundation[10] collecting and studying data on the world's surface temperature. He even used temperature records done by Benjamin Franklin and Thomas Jefferson. Muller's research examined two chief criticisms expounded by skeptics, that weather stations are unreliable and that cities create heat islands that skew the data. He found both to have no merit.[11] He concluded that the Earth has warmed 1.6 degrees Fahrenheit since the 1950s, virtually identical to what mainstream climate scientists are saying.

Putting these facts aside, there are those who would have you believe our economy would suffer if we demanded clean energy and fuel-efficient cars. Ohio faced this same situation last century when Cleveland businesses became infamous for polluting the Cuyahoga River. At the time, government regulations were virtually nonexistent. Business had free reign in disposing of their waste as they saw fit. Powerful owners and the politicians who worked for them repeatedly said that forcing them to clean up their act would put them out of business. They got their way for

Cuyahoga River Fire 1936

over a century. The result? The Cuyahoga River caught fire ten times, in 1868, 1883, 1887, 1912, 1922, 1936, 1941, 1948, 1952, and in 1969.

"No Visible Life. Some River! Chocolate-brown, oily, bubbling with subsurface gases, it oozes rather than flows. Anyone who falls into the Cuyahoga does not drown, he decays," a Cleveland citizen notes. *The Federal Water Pollution Control Administration reported, "The lower Cuyahoga has no visible signs of life, not even low forms such as leeches and sludge worms that usually thrive on wastes. It is also -- literally -- a fire hazard."*

1969 Time Magazine

A few weeks later, the oil-slicked river burst

Cuyahoga River Fire 1952

---

10 Charles and David Koch run a large privately held energy company founded by their father and both are major financiers of skeptic groups and the conservative Tea Party. Their dollars paid Fox News to market the so-called movement. The Tea Party wouldn't exist without them.

11 10/31/2011: Global warming is real skeptic's study reports, Wire Service, Arizona Republic

into flames and burned with such intensity that two railroad bridges spanning it were nearly destroyed. *"What a terrible reflection on our city,"* Cleveland Mayor Carl Stokes is reported to have said.

Cuyahoga River Fire 1952

Only after the 1969 fire was something finally done and it took the power of government by the people, for the people, to do it. The voice of powerful industrialists can still be heard today loader than ever and their money saturates our politics at the highest levels. The highest court in the land has ruled that corporations are people. I will believe that when Texas executes a corporation.

Cuyahoga River Fire

The Cuyahoga River fire was the final straw that broke the camel's back. In the summer of 1970, President Richard Nixon submitted the Reorganization Plan Number Three to Congress, which called for a single entity to govern the United States' environmental policy and thus,

the Environmental Protection Agency (EPA) was born. The EPA consolidated a variety of federal research, monitoring, standard-setting and enforcement activities into one agency. For example, the EPA assumed responsibility for monitoring air pollution, water hygiene and waste management from the Department of Health, Education and Welfare and water quality and pesticide research from the Department of the Interior. Misplaced environmental programs were finally unified under a single agency. Since its inception, the EPA has worked for a cleaner, healthier environment for the American people and the world.

Cuyahoga River Today

Many clamor for a return to the good ol' days of limited, even zero, governmental interference. This is reckless and suicidal. We cannot depend on a business, big or small, to self-regulate. History has shown the danger of doing this time after time. There must be balance between governmental oversight and economic freedom. Regardless of our fiscal fate, we must look past the end of our noses and straight into the eyes of our children and grandchildren. In a hundred years, will they curse us, or praise us?

By the time of the last Cuyahoga River fire, farmers in the United States were using large quantities of pesticides, primarily DDT. Lush green crops in the fields gave a false impression

19

of healthy, vibrant agriculture but was in fact a toxic concoction of chemicals destroying the ecosystem. Without knowing the true dangers of DDT, we used it in many ways.

When I was a boy growing up in rural

Colorado, I remember the truck moving slowly up my street pulling a trailer with the mosquito fogger mounted on it. Billowing clouds of DDT laced vapor spread out from it and went throughout my neighborhood where it lingered

and settled on every flat surface. I'm sure that I would instantly remember the smell.

We learned the hard way that DDT was toxic not only to insects and worms, but to fish and crabs. Birds that fed upon these dead insects were soon at risk. Even in small amounts, DDT would have disastrous effects on the reproduction of the animals. Even so, it wasn't until the situation had become a threat to human health that something was done.

The book *Silent Spring* published in 1962 catalyzed the environmental movement. Written by the naturalist Rachel Carson (1907-1964), the book documented in an emotional and personal way the detrimental effects of DDT. DDT is a synthetic pesticide that decimates wildlife, especially birds, which is why she named the book *Silent Spring*.

The pesticide manufacturers claimed the minute amounts found in the environment couldn't possibly be killing the birds. However, some experimental work demonstrated that even small amounts affect the reproduction of some species. More important, research demonstrated that, although concentrations were very low in the soil, atmosphere and water, concentrations were higher in plants, higher still in herbivores, and still higher as one moved up the food chain.

Rachel Carson was one of the most respected and popular science writers of her time. She completed *Silent Spring* while dying of breast cancer. The story she told and the path her life took shaped a powerful social movement that I want to believe altered the course of history but I ask you, what has really changed? We use a plethora of chemicals today with very few regulations or controls in place.

Reality Check

It is a simple fact that your life depends on individuals you will never meet. You have no choice. An army of people and machines produce everything for you, the water you drink, the food you eat, the clothes you wear, the metals in your car, and the gas that runs it, everything. Another army delivers this bounty to you. Another army services it. Anything that threatens your supply chain, threatens your way of life and the very fabric of Western Civilization.

Our technology has taken us to a precarious new plane of existence. Our roads span continents, our planes encircle the globe, we genetically engineer enough food to feed the world and our medicines extend human life. Communication satellites let you instantly talk to anyone in the world, GPS satellites insures you're never lost, and Network satellites link billions of us in a worldwide internet. Technology has created a global economy, which means our problems are no longer just ours. Human population exerts pressure universally, contamination knows no boundary, disease doesn't stop at a nation's border, and radiation doesn't play favorites. What we do matters to the rest of the world and what they do matters to us. The Earth, for all its size and ability to absorb punishment, is a finite resource that collectively, humanity is stressing to the breaking point. Civilization is perched upon the head of a pin but what finally knocks us off isn't as important as the fact that whatever it was, it was preventable. It didn't need to happen.

Regardless of the incredible span of our technology, we're all just inches away from anarchy. You don't think so? Just turn off the electrical power to your home for a week and see what happens. No lights, no cable TV, no internet, your refrigerator doesn't work and the morning shower's cold. Lack of circulation within your home quickly grows suffocating. Now turn off the electricity to your city, your state, your country. No street lights, no gas pumps, supermarket shelves are soon stripped bare, warehouses filled with rotting food, restaurants shut down, cell phones and the internet are out of reach. Without electricity, pumps stop pumping and water stops flowing through your city's pipes making indoor plumbing useless. Noxious waste builds up and the smell becomes overwhelming. The one underlying ingredient that separates civilization from this gruesome scenario is electricity. Everything else is window dressing.

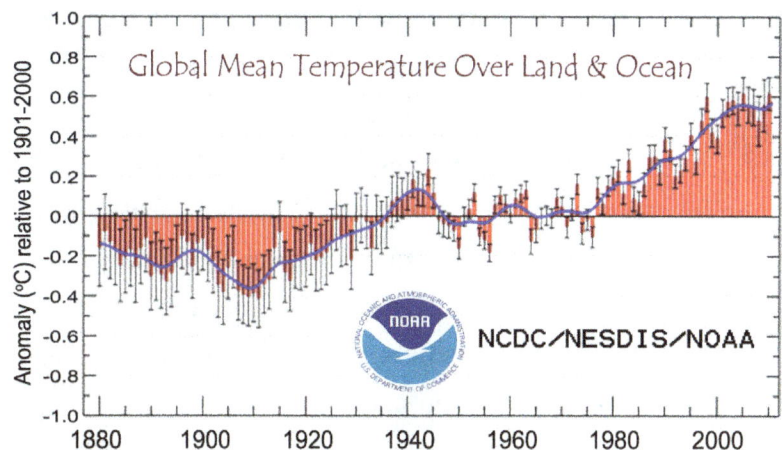

Global Mean Temperature Over Land & Ocean

NCDC/NESDIS/NOAA

However, over the last few decades, scientists have been raising the alarm that the way we are producing our electricity is changing our environment. Carbon from burning hydrocarbons is accumulating in our atmosphere helping it retain more of the sun's heat. As the temperature goes up, so does the atmosphere's ability to hold moisture. Rainstorms grow in intensity during the summer and blizzards during the winter. The oceans are also heating up and absorbing atmospheric carbon making them become more acidic. Thermal expansion of the oceans and

21

melting ice are driving sea levels up. The world's glaciers are in retreat. Storm patterns are shifting, drought stifles one region, blizzards another. After a century of unrelenting burning, 2010 was the hottest year of the hottest decade on record. Mother Earth is running a fever.[1]

It's not pretty what we're doing to our beautiful planet. Researching this book surprised even me. I hadn't realized how far we have already come. It is clear that if we do nothing and continue to careen down the road we're on, the world's ecosystems will collapse. It is only a matter of when. Earth simply cannot sustain so many people. Billions will die. All life will suffer.

Early Saturday morning on March 12, 2011, a World Wide tour bus was returning to New York City from the Mohegan Sun casino in Connecticut when the bus driver fell asleep. According to witnesses, the bus careened out of control, slammed into a guardrail, flipped on its side and skidded for over 250 yards. It stopped only after hitting a massive steel pole, cutting the top of the bus off right about seat-top level. Of the 32 passengers aboard, 14 died on impact but no one escaped unscathed.[2]

In many ways, this horrendous crash is analogous to what we are doing to our planet. We are the passengers sitting blissfully in the back of the bus while the driver falls asleep at the wheel. We need to get up and take control of the situation before humanity drives our planet over the edge. That's the purpose of this book, to make the case for the only power source capable of replacing hydrocarbons. Just remember, sunshine is the gift that keeps on giving.

---

1 Global Temperature Trends: 2010 Summation - NOAA
2 3-12-2011: Bronx tour bus crash kills at least 14 people after bus slams into pole and is split open, Bob Kappstatter, Erik Badia, Matthew Lysiak and Jonathan Lemire, DAILY NEWS WRITERS

The Elephant in the Room

The current world population requires a lot of energy just to function day-to-day, electricity to run homes and factories, gasoline for our cars, diesel to run trucks and trains, jet fuel to fly our planes, home heating oil, etc. etc. and our appetite for more is not abating. In fact, 2010 was a year of exceptionally strong growth in global energy consumption. It increased by 5.6%, the highest rate since 1973. Surpassing America for the first time, China leads the parade, increasing its usage by 11.2% taking its place as the world's largest energy consumer at 20.3% of total global energy consumption. Yet, because of their large population, China uses only one-tenth of the per capita consumption of America, so there is plenty of room to grow much larger. This pressure will affect future energy prices so you can count on paying more at the pump and the electric meter. To discuss the world's energy needs intelligently requires some effort to understand the elephantine numbers involved.

One of the biggest obstacles in understanding the enormity of our energy needs is the confusing variety of the ways we measure energy. We measure heat energy in units different from mechanical energy, different from electrical energy, different from radiant energy, etc. etc.

A convenient way to compare this menagerie is to convert them all to BTUs. BTU or British Thermal Unit is the amount of energy required to raise the heat in one pound of water 1° Fahrenheit. If you prefer to use the International System of Units, the conversion to joules or kilowatt hour (kW-hr) is easy. One of the definitions of a joule is the energy required to produce one watt of electrical power for one second.

$$Energy = c \cdot m \cdot \Delta T$$

$$1\ BTU = 1055.056\ J$$

$$\left(1 \frac{BTU}{lb \cdot °F}\right)(1\ lb \times 212° - 70°)$$

$$= 0.01759\ kW \cdot hr$$

$$J = \left(\frac{kg \cdot m^2}{s^2}\right) = W \cdot s$$

When you're talking about big numbers, what lies beyond a million? Beyond a billion? Beyond a trillion? Quadrillion! One Quadrillion BTU is 1,000,000,000,000,000 BTU and is equivalent to about 172 million barrels or 7,224,000,000 gallons of crude. This is called a Quad BTU.

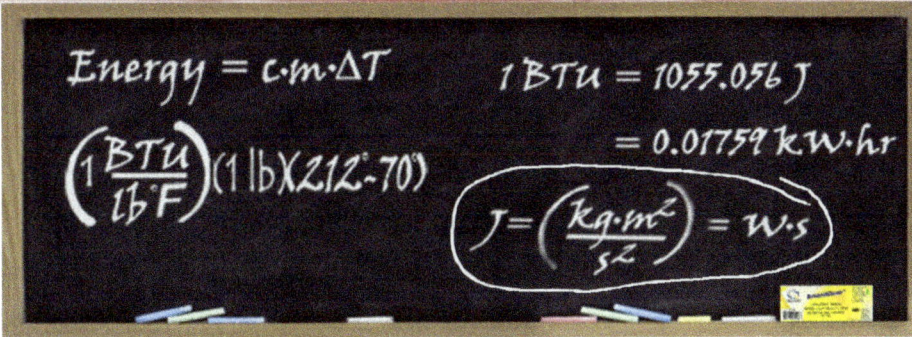

To give you a feel for using the BTU unit of energy, one gallon of water weighs 8.35 pounds, which works out to be about one pound per 16 fluid ounces. That's a cup of coffee at the local Starbucks. Using the equation provided, you will find it takes about 142 BTU to heat it.

However, what method you use to heat your coffee matters. If you used matches, lighting one right after another, you could never get where you want to go. As the temperature increases, the amount of heat loss from the liquid becomes greater than the heat gained during the burning of a single match and thus, you would never reach the desired temperature no matter how many matches you used. Any conversion has inefficiencies and losses but converting electrical energy to heat is notoriously bad. The actual energy required to heat your cup of coffee using a conventional electric heating pad is in the range of 500 BTU because of the inefficiencies involved.

This simple example endlessly multiplied creates an electrical demand that far exceeds what most people can wrap their mind around. Yet, electricity is the lifeblood of our global civilization and everyone should have at least a rough idea of the numbers involved.

What do all these big numbers mean? In 2008, the world's energy consumption was just over 500 Quad BTU per year. The U.S. Department of Energy predicts that global needs will approach 770 Quad BTU by 2035.[1] Sometime in 2010, China overtook the United States to become the world's largest energy consumer.[2] It seems the Chinese people want refrigerators, A/C, and cars. Where does all this energy come from?

Percentage of World Energy Consumption

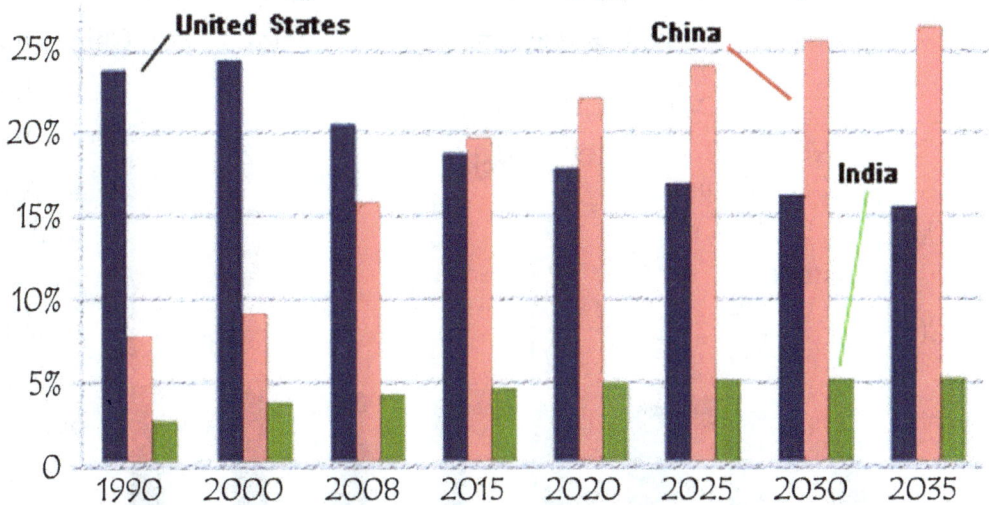

United States   China   India

25% 20% 15% 10% 5% 0

1990 2000 2008 2015 2020 2025 2030 2035

The world currently gets its energy from five basic types; Petroleum (liquid hydrocarbons), Coal (solid hydrocarbons), Natural Gas (gaseous hydrocarbons), Nuclear and a bunch of stuff everyone lumps into Renewables (hydroelectric, solar, geothermal, wind, tides, biofuels and

1 9/19/2011: EIA, International Energy Outlook 2011, Report DOE/EIA-0484(2011)
2 7/20/2010, China overtakes the United States to become world's largest energy consumer, IEA

anything else you can think of).

In 2010, worldwide crude oil consumption *increased* by 3.1% reaching a new record level

World Energy Consumption by Fuel Type

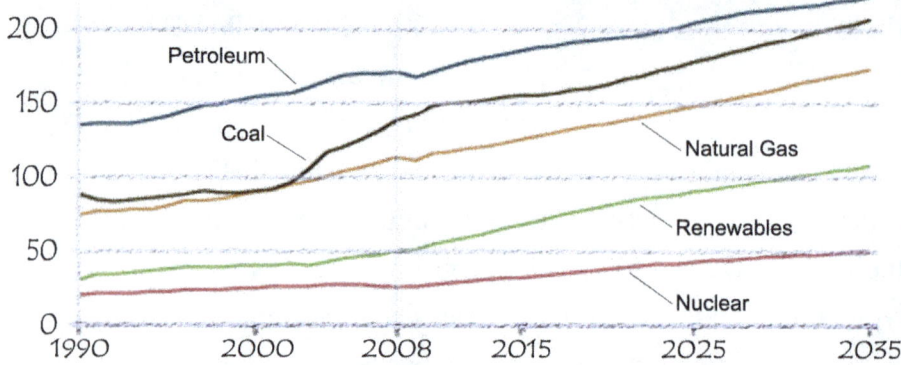

of 87.4 million barrels per day. American oil production increased for the second year in a row, and is up 11.6% since 2008.[3] This will make it hard for President Obama's political opponents to gain traction with accusations that he is blocking domestic oil production given that production went up since he took office. Many people do not realize that America is the third

World Energy Consumption

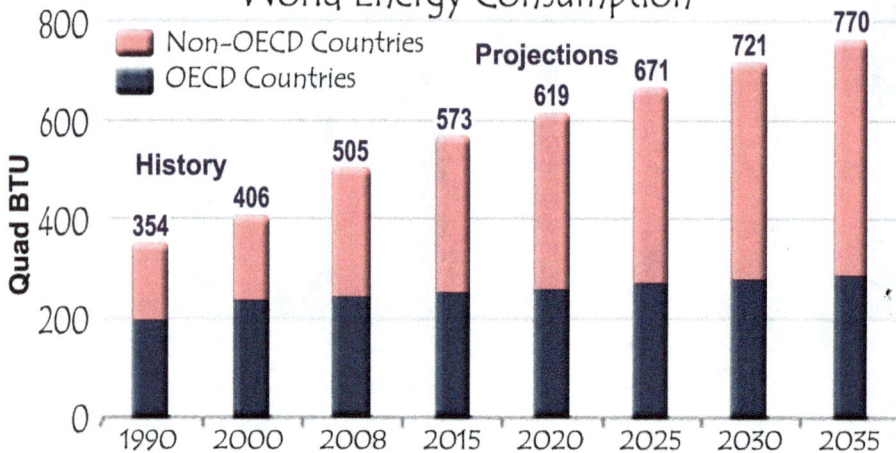

largest oil producer in the world, coming in at 8.7% of total global production. America trails only Russia (12.9%) and Saudi Arabia (12.0%).

American refining capacity fell by 0.5% in 2010 causing some to make a great deal of noise over falling refining capacity, accusing oil

companies of purposely manipulating the market to drive up profits. It is a fact that companies have shut down refineries in America and throughout Europe even while opening plants in countries with lower labor costs. Refining capacity across Europe & Eurasia declined by 1.0% while global refining capacity increased by 0.8%. Refining capacity in Africa, China, Iraq, and India respectively grew by 8.9%, 12%, 6.8%, and 3.6%. I expect the loss of American refining jobs to continue until we are importing finished products just as has happened with electronics, toys, books, washing machines and most manufacturing opportunities of the recent past.

Worldwide coal consumption increased by 7.6% in 2010, the highest growth rate since 2003. Coal's percentage of global energy consumption also increased to 29.6%, versus 25.6% 10 years ago. China consumes 48.2% of the world's coal production, almost half!

Worldwide natural gas consumption grew by a whopping 7.4% to over 120 trillion cubic feet, by far and away the largest growth rate since 1984. American consumption alone grew 21.7%. Fracking is quickly becoming our new national pastime.

Renewables grew worldwide in 2010 with China leading the way at a staggering 74.5%. In comparison, America grew by only 24.7%. Biofuel production increased by 13.8%. Wind power increased 22.7%. Hydroelectric production grew by 5.3%, breaking the previous record. China remains the top producer of hydroelectric power with 21% of the global total. These numbers sound good

3 9/19/2011, International Energy Outlook 2011, **DOE/EIA**

World Net Electricity Generation by Region

United States
Rest of OECD
— Non-OECD —
Middle East and Africa
Europe and Eurasia
Central and South America
India and Other Asia
China

trillion kilowatt-hours

(vertical axis: 0, 5, 10, 15, 20, 25, 30, 35)
(horizontal axis: 1990, 1995, 2000, 2005, 2010, 2015, 2020, 2025, 2030, 2035)

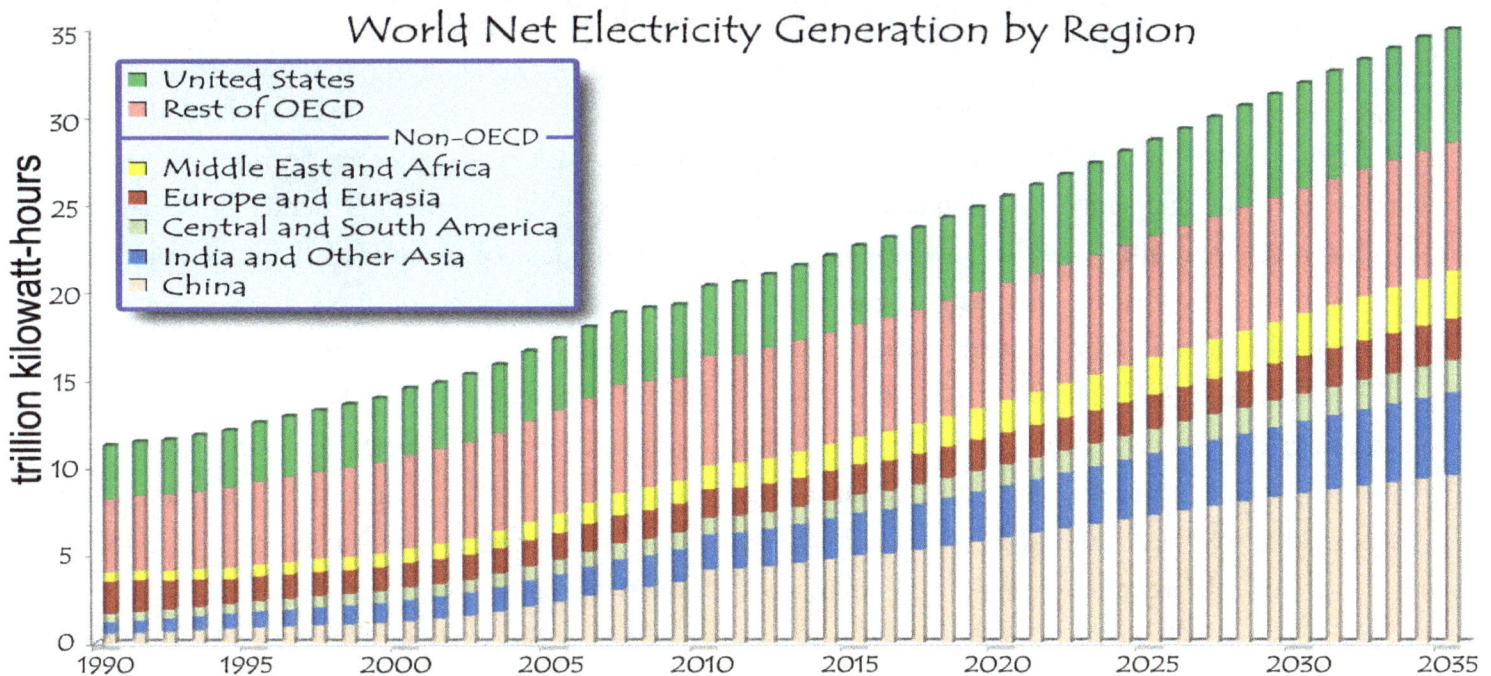

until you realize that renewables account for only 10% of global energy production and 85% of that is hydroelectric. Wind, solar and geothermal totals are literally about 1.5% of world energy production.

Global use of nuclear energy grew by 2%, a trend that I expect to reverse itself as public backlash over the aftermath of Japan's Fukushima Daiichi nuclear disaster sinks in.

There are three important things to take away from this summary. First is the fact that global energy consumption continues to rise rapidly even in the face of global climate change and dwindling raw resources. This trend is unlikely to reverse itself in the foreseeable future, assuring long-term upward pressure on energy prices. Due to the wide gap between natural gas and oil prices, I would also expect the shift toward using more natural gas to continue worldwide.

Second, the fact that China is now the world's largest consumer of energy should not be taken lightly. Their interests are not necessarily our interests, and yet they may already be the international energy corporation's most important customer. This will give China tremendous advantage, largely at the expense of the OCED countries.[4] Finally, the global increase in fossil fuel consumption supports my belief that there are only two things capable of stopping carbon emissions from rising. *One*, the world runs out of fossil fuels. We can pass all the legislation we want, write about it, talk about, and stomp our feet, but countries are still going to burn cheap energy and emit carbon dioxide by the millions of tons per day. It should not come as a surprise that global carbon emissions rose in 2010 at their fastest rate in over forty years, and it is no accident that the energy industry in general has enjoyed record profits.

Or *Two*, we give the world an energy alternative capable of replacing coal, oil, natural gas and nuclear. Only **Space Based Solar Power** has the potential to do that using technology we already have. America can lead the world into a future worth having.

4 OCED stands for Organization for Economic Co-operation and Development. Twenty countries originally signed this agreement on December 14, 1960. Since then, fourteen more countries have become members. In essence, these countries are Western Civilization (Member List)

The End of Easy Energy

In 2008, the world consumed over 500 Quad BTU. How many cups of Starbucks coffee is that? Where does humanity get all this energy day-after-day, year-after-year? Today the world has a variety of sources but coal, natural gas, and oil provides the bulk of it. These are finite resources and whether they run out today, tomorrow, or centuries from now, they will run out. Assuming our descendents are around to ask, what do you think they will say about what we are doing? Will they be indifferent to our gluttony? I doubt it.

U.S. Petroleum Production
http://www.eia.doe.gov

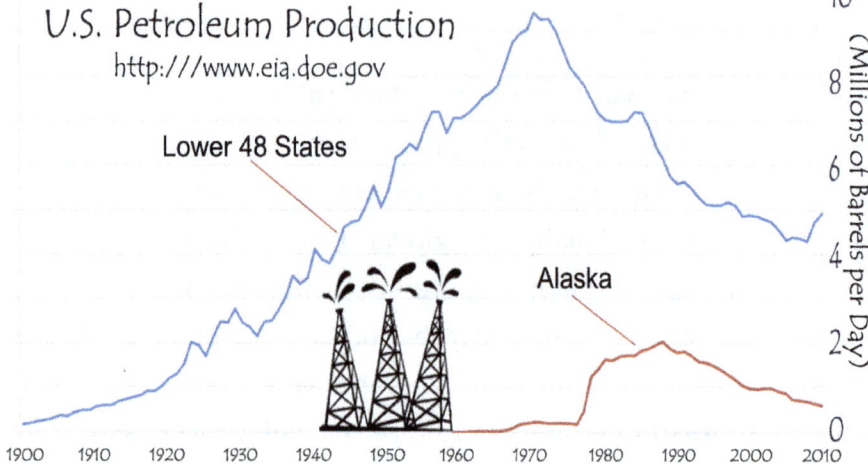

Lower 48 States

Alaska

10
8
6
4
2
0
(Millions of Barrels per Day)

1900 1910 1920 1930 1940 1950 1960 1970 1980 1990 2000 2010

There are many who say we have already found all the big oil fields Earth has to offer and without more, production will inevitably begin to fall. One of the first voices raising this alarm was Marion King Hubbert.

King Hubbert (10/5/1903 – 10/11/1989) was a geologist working at the Shell research lab in Houston, Texas in 1956 when he came up with the Hubbert Curve and Peak Theory, basic components of Peak Oil.

Peak Oil is the point in time when global petroleum extraction reaches a maximum value, after which the rate of extraction enters terminal decline. An observer will note the production rate from an oil well grows exponentially over time until it peaks and then declines, sometimes rapidly, until the well is depleted. It happens to individual oil wells and fields of related oil wells. King Hubbert applied this concept to American oil production and predicted Peak Oil would occur in the early 1970's. He was widely criticized by oil experts and economists, but fifteen years later, Hubbert's prediction came true. The graph clearly shows the bell shape of the American Hubbert curve.

Since then, Hubbert and his ideas have gained many knowledgeable followers. Colin Campbell is an Oxford-trained petroleum geologist with 40 years experience working for Texaco, British Petroleum, Amoco, Shenandoah Oil, Norsk Hydro, and Fina.[1] Using Hubbert's methods, he predicts the peak in world oil production has already happened or will within the next few years. The fierce debate over the precise date of the peak misses the point.

Whether it is yesterday, today, or a decade from now, petroleum is a finite natural resource and production will inevitably begin the long remorseless decline due to natural depletion. Because petroleum is a fundamental part of the world's economy, as production falters, correspondingly, prices will rise putting an already tight economy under increasing pressure. Many believe we have already passed the peak and started fighting over what remains. Western involvement in Middle Eastern politics is a case in point. (Note: For those with a calculus background, let your eye measure the area under

1   Dr Colin Campbell, Peak Oil, ASPO

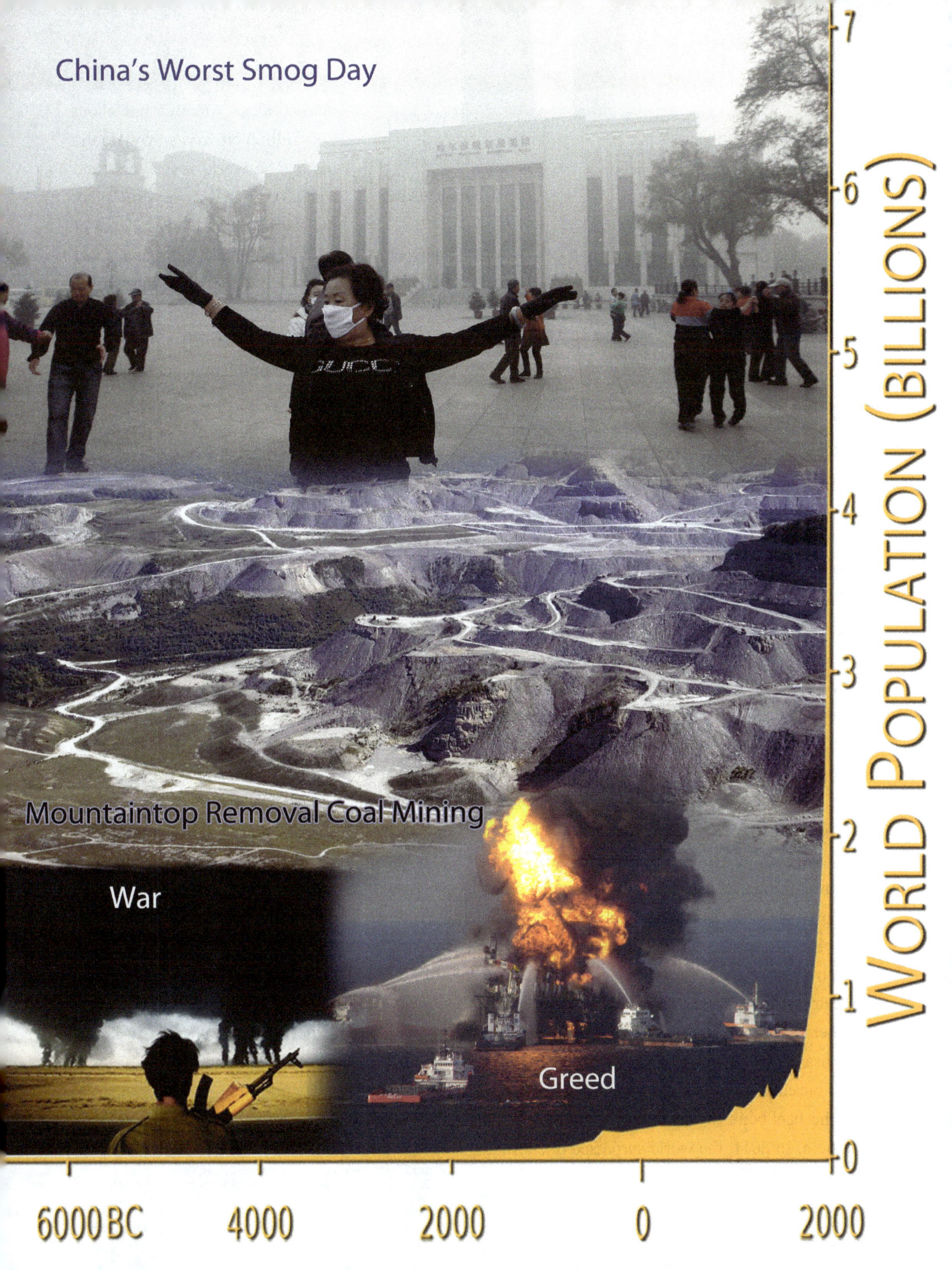

China's Worst Smog Day

Mountaintop Removal Coal Mining

War

Greed

WORLD POPULATION (BILLIONS)

7
6
5
4
3
2
1
0

6000 BC   4000   2000   0   2000

Growing Gap Between Global Oil Production and Discovery

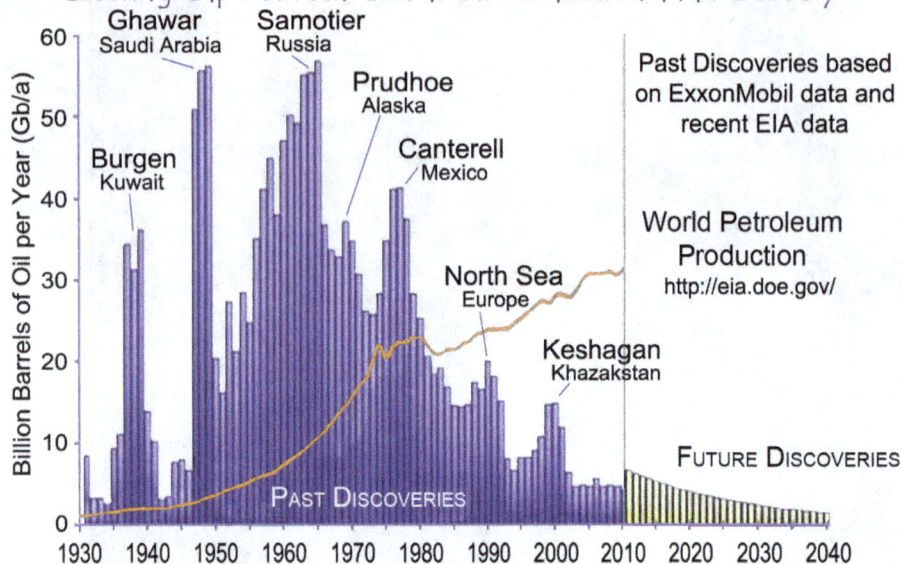

Ghawar
Saudi Arabia

Samotier
Russia

Prudhoe
Alaska

Burgen
Kuwait

Canterell
Mexico

North Sea
Europe

Past Discoveries based on ExxonMobil data and recent EIA data

World Petroleum Production
http://eia.doe.gov/

Keshagan
Khazakstan

PAST DISCOVERIES

FUTURE DISCOVERIES

Billion Barrels of Oil per Year (Gb/a)

60 50 40 30 20 10 0

1930 1940 1950 1960 1970 1980 1990 2000 2010 2020 2030 2040

the two curves in the global oil production graph. How much oil remains in the ground?)

In 1996, engineer Richard C. Duncan introduced a paper titled *The Olduvai Theory: Sliding Towards a Post-Industrial Stone Age.* In it, he correlates the ratio of world energy production with world population. It predicts the life expectancy of Industrial Civilization is less than or equal to 100 years: 1930-2030. One thing is certain, if we wait to develop an energy source capable of taking the place of existing coal, gas and nuclear power plants until we actually need it, we are going to be in big trouble.

Starting about a century ago, we began burning complex hydrocarbons, primarily coal, to produce heat that generated steam to run a turbine that produced electricity. According to the International Energy Agency (IEA) Clean Coal Centre, there are over 2300 coal fired electrical generating plants in the world. Natural gas fuels thousands more and heats hundreds of millions of households worldwide. In addition, crude oil is the backbone of our interstate transportation system. In 2011, the world surpassed one billion

automobiles.[2] America has about 300 million citizens so you do the math. The world loves cars (and trucks) as much as we do. Today the world burns hydrocarbons at a blistering rate releasing most of the byproducts into the Earth's environment with little or no regard for the long-term consequences. And the problem is getting bigger every year. Worldwide carbon-dioxide emissions are predicted to more than double over the next few decades from 16 billion tons to 38 billion tons per year.[3]

The IEA estimates there are 750 billion metric tons[4] of coal still in the ground. We currently use about 7 billion tons a year. Knowing that demand is increasing, simple math dictates there is less than 100 years of coal left. Are we so addicted to coal that we cannot stop burning it before it is all gone? What will this do to our beautiful planet?

When coal burns completely, the products are gaseous carbon dioxide (CO_2) and water (H_2O). A typical coal power plant generates almost four million tons of CO_2 a year, but the combustion is never perfect. Burning coal produces soot (small particles of impure carbon), aromatic hydrocarbons like tar (the stuff found in cigarette smoke), sulfur dioxide and nitrogen oxide[5] (the stuff that makes acid rain), carbon monoxide, mercury, arsenic, lead, cadmium, and even uranium. I don't need to be a chemist to know this *stuff* poisons our environment and yet we continue to do it day after day, year after year.

We know with a great deal of accuracy from

2 9/12/2011: World's 10 largest auto markets, CNBC
3 2004: Global Greenhouse Gas Data, Environmental Protection Agency
4 Metric ton = 1000 kg
5 Nitrogen Oxides - Pollution Prevention and Control, World Bank Group

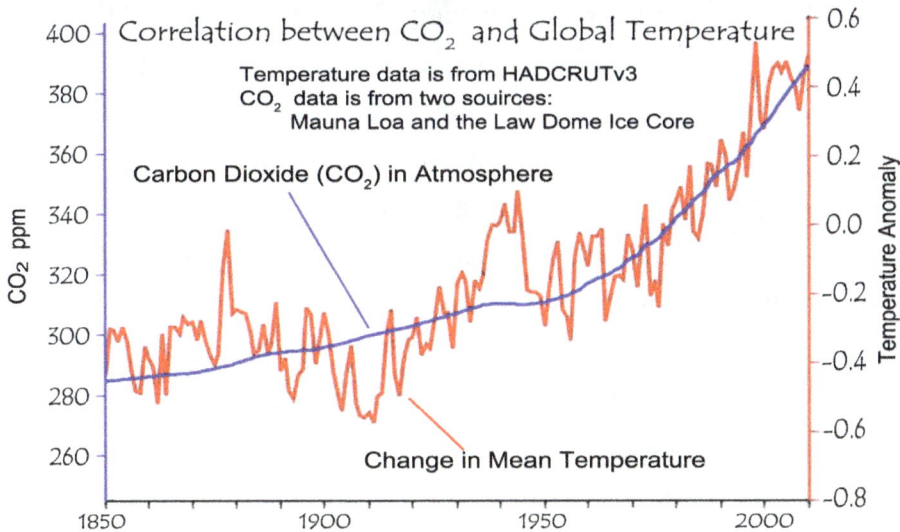

Correlation between CO₂ and Global Temperature

Temperature data is from HADCRUTv3
CO₂ data is from two sources:
Mauna Loa and the Law Dome Ice Core

Carbon Dioxide (CO₂) in Atmosphere

Change in Mean Temperature

laboratory experiments and empirical data what increasing the amount of CO_2 in our atmosphere will do. It increases the amount of heat retained by the atmosphere and warms our little world. This is the Greenhouse Effect and CO_2 is a greenhouse gas. This is also why the phenomena is called Global Warming. A better name would have been Global Climate Change. Regardless of what we call it, it threatens the entire world. Even if CO_2 emissions were all we had to worry about, it would still constitute a major reason to stop burning hydrocarbons to produce our electricity.

Scientists, philosophers and various legal scholars recently got together and brainstormed ways we could artificially cool the Earth. They came up with ideas ranging from painting everything white to positioning sunshades in space between the sun and us. Their conclusion? Geo-engineering is a risky way to combat global warming and is not a viable alternative to reducing carbon emissions.[6] There is no magic bullet here. We must kick the burning habit just like a twenty-year smoker quits cigarettes one urge at a time, or suffer an early death.

Burning coal also produces a toxic ash or sludge containing one or more of the following substances in quantities from parts-per-million (ppm) to several percent: arsenic, beryllium, boron, cadmium, chromium, chromium VI, cobalt, lead, manganese, mercury, molybdenum, selenium, strontium, thallium, and vanadium, along with dioxins and poly-aromatic hydrocarbons. The specific constituents depend upon what kind of coal you're burning. Anthracite is the best, bituminous is what you burn when you can't get anthracite, and lignite is not much better than peat. They're all carbon-dirty. There is no such thing as clean coal.

On October 11, 2000 the bottom of a huge coal sludge pond owned by Massey Energy in Martin County, Kentucky, collapsed into an abandoned mine below and escaped out of the mine openings. Martin County Sludge Spill released 306 million gallons upon the residents of Kentucky and West Virginia. They awoke to find black sludge over five feet deep in places polluting hundreds of miles of the Big Sandy River, Tug Fork River, tributaries and eventually, the Ohio River. It exterminated all aquatic life in Coldwater Fork and Wolf Creek and contaminated the water supply for over 27,000 residents. The EPA ranks this as one of the worst environmental disasters ever in southeastern America.[7] Have you ever heard of it? Not many have.

As bad as that was, it wasn't the worst sludge accident to date. On December 22, 2008, the Kingston Sludge Spill released 1.1 billion gallons of coal fly ash slurry when a containment dike ruptured at the Tennessee Valley Authority's Kingston Fossil Plant in Roane County, Tennessee. It buried twelve homes and damaged

6 12/4/2011: Geo-engineering a risky way to combat warming,
   AP, Arthur Max

7 1/10/2001: Martin County Coal Slurry Spill, EPA

forty more. The mixture of fly ash and water laced with arsenic and heavy metals polluted the Emory River and Clinch River flowing downstream all the way to the Tennessee River.[8] It was the largest fly ash discharge in United States history. Let's hope it was the last.

Not! The latest in a long list of smaller disasters happened on November 1, 2011. A land collapse at a coal power plant in Oak Creek, Wisconsin dumped a small hill's worth of coal slurry into Lake Michigan. Like similar spills in Tennessee and elsewhere, this event deposited untold amounts of arsenic, mercury, heavy metals, and carbon compounds into one of America's largest sources of fresh water.[9] Ironically, just two weeks before, the House of Representatives passed a bill that would remove most of the EPA's authority to control coal ash dumps. Oak Creek is in Wisconsin's 1st District, represented by Paul Ryan at the time of this writing. I wonder if he's still in favor of eliminating the EPA? Who does Mr. Paul Ryan think should regulate toxins? We need our government to watch these companies to make sure they do what's necessary and when that fails, to clean up the mess afterwards and hold them accountable.

Mining coal has its own unique problems. At 3:27 PM, on April 5, 2010, Massey Energy's Upper Big Branch coal mine exploded killing twenty-nine West Virginia miners. A state-funded independent investigation would later put the blame squarely on the owner of the mine, Massey Energy, concluding that it had *"made life difficult"* for miners who tried to address safety and built *"a culture in which wrongdoing became*

acceptable."[10] January 2, 2006, the Sago Mine exploded and trapped thirteen miners for nearly two days. Only one survived. The reason? A series of bad decisions made by the mine's owner and allowed by the state and federal agencies charged with mine safety. They gutted existing regulations just to pad their bottom line. The result? The death of twelve men. On September 23, 2001, a Jim Walter Resources coal mine in Brookwood, Alabama, exploded killing thirteen miners.[11] On November 20, 1968, methane sparked at the Consol No. 9 coal mine in West Virginia with ninety-nine miners inside. The explosion was large enough to be felt twelve miles away in Fairmont. Twenty-one men escaped, rescue teams recovered fifty-nine bodies and the mine still entombs nineteen miners. The cause of the explosion was never determined. These few instances are but a scratch on the surface of a seemingly endless list of bad decisions.

From January 2001 to October 2004, the records show that coal mining was the most deadly job in China, averaging a death per week.[12] In 2003, the average Chinese coal miner produced 321 tons of coal a year compared to 14,500 tons by his American counterpart. However, for every 100 tons of coal they mine, the death rate for Chinese miners is 100 times that of American miners. Could it be that the self-regulation built into the communist system is inherently flawed?

We have learned over many years of pain and tragedy that human endeavors, from banking to coal mining, must be actively and efficiently regulated to ensure safe and fair operating conditions, especially in situations where profit can trump human decency and end in death

---

8 12/29/2008: Rivers high in arsenic, heavy metals after sludge spill, CNN

9 11/1/2011: Another Coal Ash Spill – This Time in Lake Michigan, Sierra Club

10 5/19/2011: Report on the Upper Big Branch Mine Explosion, NYT

11 Jim Walters Mine Disaster, United Mine Workers of America

12 11/13/2004: Coal mining - Most deadly job in China

and/or bankruptcy. Bringing home the bacon shouldn't cost anyone their life. Here in America we have learned that an independent inspection agency monitoring and measuring a company's endeavors is the only solution and the federal government is best suited to do that, certainly not the company itself. Self-regulation has proven disastrous in the past and must be a lesson we never forget.

The problem is this; every administration appoints thousands of people to powerful regulatory positions within the federal government. It is what they do. It is how a president puts their personal mark on an administration. Some presidents appoint people loyal to them but with very little knowledge of the industry they are regulating. That can lead to disaster through ignorance. Katrina and FEMA jump to mind. The other option is to look to the industry itself to find qualified individuals but when you appoint corporate insiders to these positions, the decisions they make may not always be in the best interests of the workers or the environment. In extreme cases, the inspection agency becomes a toothless sham leading to disaster through arrogance. The poster child of this is the BP Gulf Disaster and the Minerals Management Service, the government agency responsible for policing offshore drilling. The agency had become totally ineffective, a rubber stamp.[13] We need our government to do its job.

As dangerous as underground coal mining is, strip mining, or Mountaintop Removal Mining, is even more destructive to the environment. It has already wiped out thousands of miles of streams and threatens another 1.4 million acres in the Appalachia Mountains.[14] Using this technique, mining companies routinely blow the tops off mountains enabling them to extract many tons of coal every few seconds. One truck holds almost 450 metric tons of coal. Coal dust, if left untreated, will accumulate on machinery and must be constantly washed off using a tremendous amount of water. To minimize disposal costs, they dump millions of tons of waste rock into the valleys below, poisoning drinking water, destroying wildlife habitat, and increasing the risk of flooding. The practice has wiped out entire towns, forcing people to abandon their homes and businesses. Residents in West Virginia, only a few miles from its capital, Charleston, cannot drink tap water. It is full of arsenic, barium, lead, manganese and other chemicals at concentrations federal regulators say cause cancer and damage the kidneys and nervous system.[15] It was directly traced back to coal mining.

From experience, we know that mining coal destroys the environment around the mine, burning coal sends pollutants into the atmosphere where they can damage the environment for hundreds or even thousands of miles away, and the toxic residue left after coal is burned must be safely stored for many years. This is only a glimpse of the baggage coal brings to the table. It doesn't take a genius to see we must find another way to generate electricity, something clean and environmentally friendly

What about natural gas? While natural gas power plants do not generate fly ash, they do produce nitrous oxides and carbon dioxide (CO_2) in huge quantities. It is about half that of coal, but when you start with such a dirty process, cutting it in half just isn't good enough. At best, natural gas

13 4/17/2011: Regulation of Offshore Rigs is a Work in Progress
14 EPA, Mid-Atlantic Mountaintop Mining

15 9/12/2009: Clean Water Laws Are Neglected at a Cost in Suffering, NY Times

Roughly 200 tanker trucks deliver water for the fracturing process.

A pumper truck injects a mix of sand, water and chemicals into the well.

Natural gas flows out of well.

Recovered water is stored in open pits, then taken to a treatment plant.

Storage tanks

Natural gas is trucked to a pipeline for delivery.

0 Feet

Water table   Well

1,000

Hydraulic Fracturing

Hydraulic fracturing, or "fracing," involves the injection of more than a million gallons of water, sand and chemicals at high pressure down and across into horizontally drilled wells as far as 10,000 feet below the surface. The pressurized mixture causes the rock layer, in this case the Marcellus Shale, to crack. These fissures are held open by the sand particles so that natural gas from the shale can flow up the well.

2,000

3,000

4,000

5,000

6,000

7,000

Sand keeps fissures open

Natural gas flows from fissures into well

Shale

Fissure

Well

Mixture of water, sand and chemical agents

Fissures

Well turns horizontal

Marcellus Shale

The shale is fractured by the pressure inside the well.

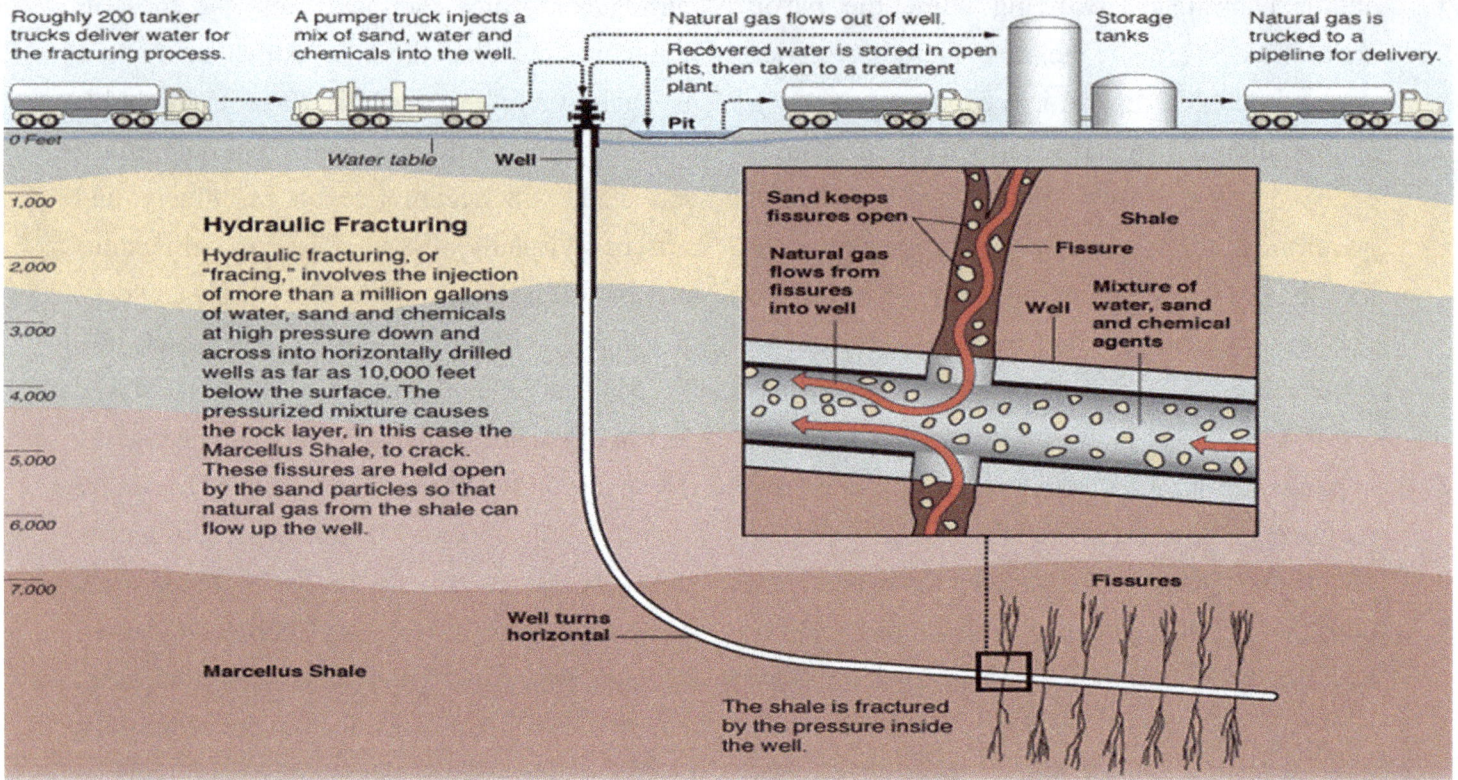

Graphic by Al Granberg

can buy time to build the infrastructure needed to implement **SBSP**. Burning natural gas is simply trading one monster for another that kills you slower. It is not the final answer.

To maximise the extraction of natural gas, the industry uses a process called Hydraulic Fracturing, or fracking for short. The drilling industry prefers spelling it without the k, fracing. Regardless of how you spell it, fracking is a process that fractures the rocks deep underground to increase the rate and ultimate recovery of oil and natural gas. *Injecting* a high-pressure chemical concoction into the well opens the fractures and causes them to spread through the rock. They add sand and other material to prevent the fractures from closing when the *injection* phase stops. The pressure is so great and the fractures so destructive, that it appears fracking causes earthquakes.[16] I'm not talking about one well here. From a total of about 500,000 gas wells in America,[17] 26,000 are known fracked wells spread out in 16 states.[18] Just imagine what happens when we frack them all!

But earthquakes are the least of our worries when it comes to fracking. The process has taken on an ominous tone with images of flames shooting out of a kitchen faucet and toxic bubbles killing wildlife along rural waterways. At this point it looks like fracking releases methane into our aquifers contaminating our drinking water. The chemical concoction used in the process may also contribute to the contamination but what exactly they are, we don't know.

The industry refuses to tell us what chemicals they use. The secrecy surrounding fracking does nothing to calm the fears of the people who must live with the consequences. The public needs to know what chemicals they use. Instead, the corporations that run these wells have been given a political pass on our most basic environmental

16 1/1/12: Fracking debated as possible quake link, Wire Service, Arizona Republic

17 Distribution of Wells 2009
18 5/6/2011: Frack and Ruin – The rise of hydraulic fracking,

133rd Ave NW
Williston, ND 58801
48.459832, -103.559058

In this image are thirteen of the thousands of wells and drilling infrastructure in North Dakota. These are spreadout among farms and close to surface water reservoirs.

Lone Tree Lake

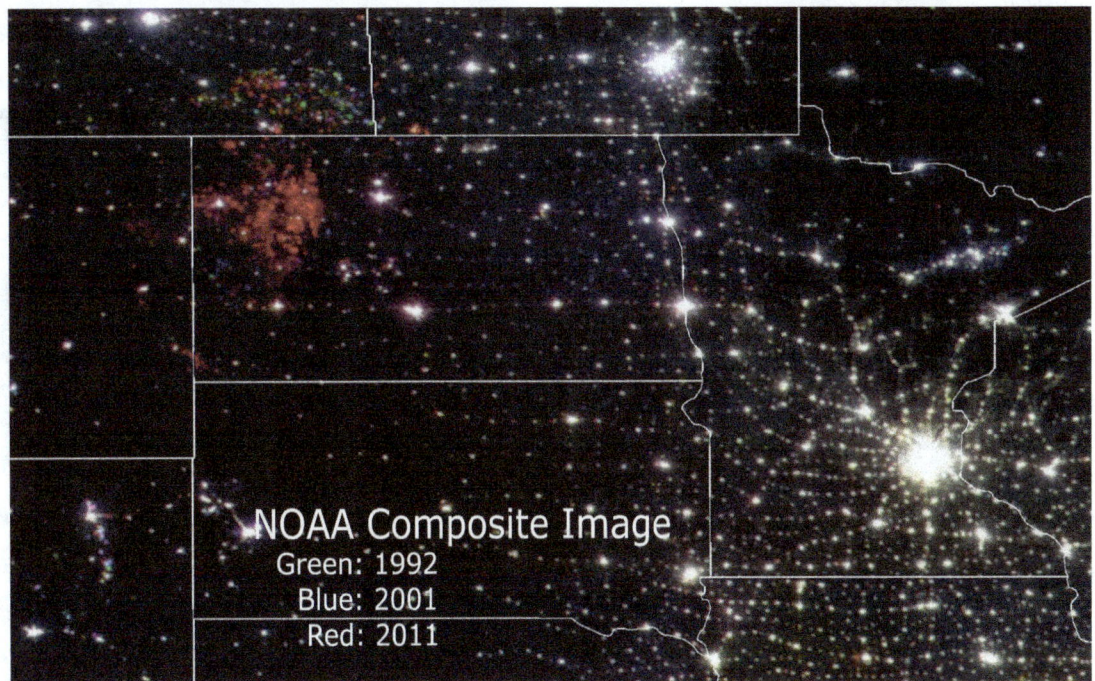

NOAA Composite Image
Green: 1992
Blue: 2001
Red: 2011

safeguards. They operate without governmental supervision or regulations. I sure hope they know what they're doing, don't you?

The Energy Policy Act of 2005[19] exempted fracked wells from being classified as *injection* wells. This excluded them from federal regulation under the Safe Drinking Water Act. Accusations of groundwater contamination have brought into question whether the exemption is appropriate. I think the answer is looking you in the face. These are *injection* wells and need to comply with existing law.

The larger image shows a small portion of the huge gas field in North Dakota. To show the scale of this gas field, NASA put together a composite image using three photos taken from NOAA satellites in 1992, 2001 and 2011. Lights in the three photos were artificially colored red, green and blue (RGB) using filters and then combined into one image. These three colors combine to show the towns and cities that was there prior to 1992 as white light. The sprinkling of blue and green light cannot hold a candle to the broad swath of red lights that has appeared since 2001. Running night and day, the extraordinary large

19 Energy Policy Act of 2005; Public Law 109-58-Aug 8, 2005

number of drilling rigs can be seen from space.

Two studies released in 2009, one by the U.S. Department of Energy and the other by the Ground Water Protection Council, address hydraulic fracturing safety concerns. They suggested upgrading the industry regulations including releasing a list of chemicals used in the process. Chemicals used in the fracturing fluid include kerosene, benzene, toluene, xylene, and formaldehyde but this list is sorely incomplete. Why do we put up with this? For cheap energy? There must be a better way!

In June 2010, the Wyoming Oil and Gas Conservation Commission voted to require full disclosure of the hydraulic fracturing fluids used in natural gas exploration. By August 2011, they changed their mind and agreed to let the drilling companies keep the hundreds of chemicals used in fracking a secret. I wonder what happened.

In late April 2011, Chesapeake Energy Corp lost control of a natural gas well spilling thousands of gallons of chemicals into the surrounding environment and forcing the evacuation of nearby families.[20] In December 2011, the EPA finally admitted for the first time that fracking may cause groundwater contamination.[21] A month later, Bulgaria became the first nation to ban fracking permanently.[22]

Natural gas is essentially methane, a greenhouse gas over twenty times as potent as carbon dioxide. You see it being burned off oil wells, refineries, and gas liquefaction plants.

Methane Measurements Global Average

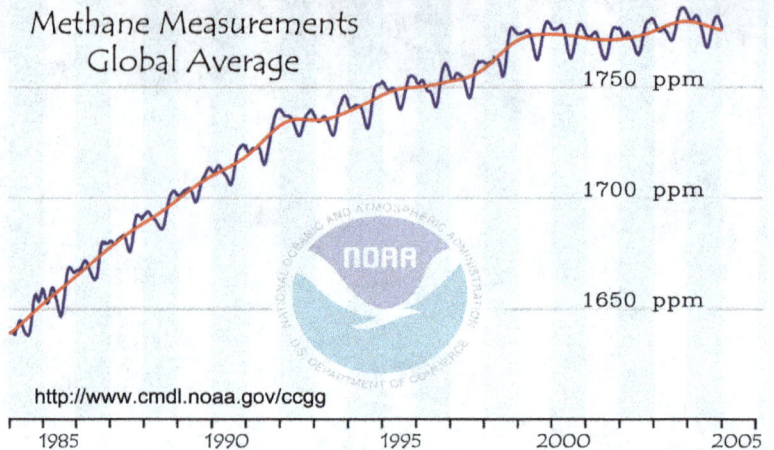

http://www.cmdl.noaa.gov/ccgg

1750 ppm
1700 ppm
1650 ppm

1985   1990   1995   2000   2005

The industry believes, and rightly so, that it is the lesser of two evils to introduce carbon dioxide into the atmosphere rather than raw methane. Despite all this effort, the percentage of methane is increasing in our atmosphere. Measurements in the arctic reached 1,850 ppm in 2010, the highest

its been for the last 400,000 years.

On a purely human scale, our planet seems like a big place but there are limits on what it can absorb. Something relentlessly repeated day after day, year after year, without pause will eventually change our environment. At the very least, we are giving our world a fever. At worst, we are poisoning the ground we walk on and the air we breathe for the sake of easy energy.

Call the men in the white coats to come take me away. I hear the angry voices of my granddaughter's granddaughter echoing through my head, cursing me for what I didn't do. I didn't stop the madness as humanity careened towards the edge of darkness.

20 4/21/2011:Gas well blowout spews chemical laced water, Wire Service, Arizona Republic
21 12/9/2011: EPA suggests fracking-pollution link, AP, Mead Gruver
22 1/19/2012: Bulgaria bans 'fracking' in gas, oil exploration, Wire Service, Arizona Republic

Pandora's Box

Nuclear fission splits the nucleus of an atom into smaller atoms. Nuclear fusion does the opposite. It joins two or more atoms together to form a single heavier atom. Both release huge amounts of energy in the process.

Since the 1950s, nuclear fusion has only been a few years away from delivering cheap clean energy to a hungry world inspiring the investment of billions of dollars. Despite all this effort, the technology has eluded us but we keep on trying. Its latest multibillion-dollar incarnation is the International Thermonuclear Experimental Reactor or ITER at Cadarache, France. The ITER is an international research project whose goal is to bring the world's first full-scale electricity-producing fusion power plant online by 2018.[1] If all goes to plan, it will be the first fusion reactor to reliably produce more energy than is put in. I'm not holding my breath. I've heard this story before.

Seven countries shared the initial $13,000,000,000 price tag, the European Union (EU), India, Japan, China, Russia, South Korea and the United States. The EU, as host country, contributed 45% of the cost, with the other six countries contributing 9% each. Begun in 2006,

the ITER project costs have spiraled upward. Construction costs alone have tripled. Nuclear fusion continues to suck up a lot of research money for very little return on the investment. If this had been directed towards **SBSP**, our kids would be dreaming of the high frontier instead of worrying about nuclear proliferation and burning coal polluting our air.

Unlike fusion, fissile materials are those that split when bombarded by neutrons in a self-sustaining chain-reaction releasing enormous amounts of energy in the process. In nuclear reactors, fission is controlled and the heat energy harnessed to produce electricity. In nuclear weapons, the energy releases all at once producing a violent explosion. The most important fissile materials for nuclear energy and nuclear weapons are plutonium-239 and uranium-235. Uranium-235 occurs in nature, plutonium-239 does not.

After massive military research, we have fission technology down pretty well. Not only does it make a great bomb, we use fission to heat water to turn electrical generating turbines that power America. America is the world's largest user of these nuclear power plants. Our country's 104 reactors supplied over 20% of our total electrical output in 2010. Yet, there hasn't been a new nuclear power plant started in America for the last thirty years. Why? The short answer is that nuclear plants are too costly to build and the price of insurance has recently gone through the roof. Seriously. Since the Fukushima disaster on March 11, 2011, you cannot find an insurance company willing to risk it. All around the world, countries are backing off nuclear energy.

Four months after the disaster, the Japanese Prime Minister Naoto Kan said that Japan would decrease and eventually end its reliance on nuclear energy.[2] *"We will aim to bring about a society that can exist without nuclear power."* Twelve months after the disaster, all but one of Japan's 54 nuclear reactors was off line. Japan has turned to oil and coal to make up for the shortfall and businesses have been required to conserve until it hurts. The last reactor is scheduled to be shut down in May 2012.[3]

China, the world's third largest economy, announced they were revisiting their long-range nuclear plan.[4] They also announced they were increasing their target for solar power capacity from the previous 20 GW to 50 GW. Using conventional ground based solar cells would require a solar array of over 4200 km^2. They have something else in mind. Speaking about China's ambitious space solar energy program, 90-year-old Chinese space technology pioneer Wang Xiji said **SBSP** would promote international cooperation. *"Whoever takes the lead in the development and utilization of clean and renewable energy, and the space and aviation industry, will be the world leader."*[5]

German Chancellor Angela Merkel called for a measured exit from nuclear power and an expedited transition to renewable energy. *"When the apparently impossible happens in such a highly developed country as Japan, then the whole situation changes."* A month after Fukushima, Germany shut down the eight oldest nuclear reactors and plans to replace the remaining nine by 2022.[6] They are turning away from the beast.

2   7/14/2011: Japanese leader calls for an end to nuclear energy, Wire Service, Arizona Republic
3   3/27/2012: Japan has shut down another nuke reactor, Wire Service, AZ Rep
4   3/31/2011: China boosts solar and cuts nuclear following Fukushima crisis – BusinessGreen
5   9/2/2011: China unveils plans for solar power station in space
6   9/15/2011: Germany shuts down eight nuclear reactors, Wire Service, AZ Rep

On October 30, 2011, work came to a halt on the 2 GW nuclear plant under construction in an earthquake prone area of Bulgaria on the Danube River. It seems that no one will lend the money to complete the project except for the contractor, Russian state-run Rosatom Corporation. After watching one of the richest nations on Earth struggle with Fukushima on global TV, the EU's poorest nation was looking for any excuse to back out of the deal without totally pissing everyone off. By everyone, I mean the Russians.

Here in America we have also built nuclear reactors on major fault lines, daring Mother Nature to do her worst. Diablo Canyon Power Plant in California is located on or near at least a dozen earthquake faults. The Shoreline Fault is a mere 1,500 feet from the base of reactor No. 1. Other faults are close by, including the area where the 2.5-million-gallon cooling water reservoir is located. Both the core and the spent fuel rod storage pools need large amounts of water to prevent meltdown.

No one thinks about the East Coast when worrying about earthquakes but on August 23, 2011, a 5.9-magnitude quake centered on Louisa, Virginia, rattled the entire Eastern Seaboard. The shaking was dramatic in Washington, D.C., and felt to some degree all the way to Boston. As the ground shook, the two nuclear reactors at the North Anna Power Station automatically began shutting down. Even though nothing was damaged, it took over a month to bring the power station back online.

Nuclear energy goes far beyond being dangerous, yet, for many Americans, it remains a viable alternative. Let me remind them of our own brush with disaster. The morning of March 28, 1979 started out like any other in the Three Mile

Island Nuclear Generating Station in Dauphin County, Pennsylvania. Then in the early morning hours, failures in the non-nuclear secondary system closely followed by a stuck valve in the primary system, allowed large amounts of nuclear reactor coolant to escape into the environment as steam. Plant operators compounded the initial failure, mistakenly believing that too much coolant water in the reactor was causing the steam pressure release. They convinced the Nuclear Regulatory Commission (NRC) to authorize the release of 40,000 gallons of radioactive coolant water directly into the Susquehanna River. By the time they realized their mistake, a meltdown was narrowly averted.

Meltdown was not averted seventeen years later when on April 26th, 1986, a terrible accident occurred at the Chernobyl Nuclear Power Plant in what is now the Ukraine. The disaster began during a systems test when a sudden power surge in Reactor Number Four triggered an emergency shutdown that generated an even larger spike leading to an explosion. The blast opened up a hole in the containment dome and exposed the graphite control rods to air, causing them to ignite. The resulting plume of radioactive fallout entered the atmosphere and drifted over an extensive geographical area, including the nearby city of Pripyat, large parts of the western Soviet Union, Eastern Europe, Western Europe, and Northern Europe. The USSR abandoned large areas in Ukraine, Belarus, and Russia and forced the resettlement of over 336,000 people. According to official post-Soviet data, about 60% of the fallout landed in Belarus. This remains one of the worst nuclear power plant accidents in history, and one of two incidents classified as a Level 7 event on the International Nuclear Event Scale.

This generation will never forget the other Level 7 event. It began on March 9, 2011, when a 7.2 magnitude earthquake shook Japan's east coast of Tōhoku causing light damage. Two days later, a 9.0 magnitude megathrust earthquake rocked the Oshika Peninsula. It was the strongest quake to hit Japan in human memory and the fourth largest since seismological record keeping began in 1897. It created a tsunami wave over 10 meters (33 ft) high. The disaster left tens of thousands dead and destroyed buildings and infrastructure. However, more to the point, the one-two punch severely damaged one of Japan's major nuclear power stations.

Constructed in the 1970s right on Japan's northeast coast, Fukushima I Nuclear Power Plant consists of six light water reactors driving electrical generators with a combined power of 4.7 GW. Fukushima was one of the largest nuclear power stations in the world. The earthquake and subsequent tsunami disabled the reactor cooling systems of four reactors causing repeated explosions and triggering widespread evacuation surrounding the plant. Classified as a Level 7 incident, it was and still is four simultaneous disasters at the same time. For many months afterward, meltdown continued to threaten a huge swath of Japan's northern territory, far more than the natural disaster had affected. Finally, on September 29, 2011, 6½ months after the tsunami, temperature levels in the cooling pools of all four reactors fell below the boiling point for the first time since March 11. Even so, Fukushima remains in critical condition as of this writing. Japan released a report in November 2011, predicting it will take thirty years or more to safely decommission Fukushima. Ultimately, a concrete and steel sarcophagus will encase it.

It is grotesquely ironic that Japan should face this disaster since they are the only nation on Earth that has endured nuclear weapons, Hiroshima on August 6, 1945, and Nagasaki three days later. *If radiation is so dangerous, why do people live in these cities today?* There are two ways to produce radioactivity from an atomic blast. The first is from fallout of the nuclear material itself, uranium or plutonium particles. The Hiroshima A-bomb used uranium whereas the Nagasaki A-bomb used plutonium. We have all seen the images, the huge mushroom cloud rising above the city like Godzilla in a bad movie. Both bombs exploded above ground zero at altitudes of 600 meters and 500 meters, respectively, formed huge fireballs that rose high up in the stratosphere, and dispersed its load of radioactive material around the world. Scientists have calculated that about 10% of the nuclear material in the bombs underwent fission and the remaining 90% disappeared into our global environment as radioactive fallout. Currently, the radioactivity in the two cities is difficult to distinguish from the amounts deposited during atmospheric atomic bomb tests conducted in the 1950s and 1960s.

The second way to produce radioactivity is by neutron activation. Due to the physics involved, A-bomb radiation is 90% gamma rays and 10% neutrons. Neutron bombardment causes ordinary, non-radioactive materials to become radioactive. Gamma rays do not. It is the fact that the bombs were detonated so far above ground that they could not produce the degree of contamination people associate with nuclear test sites such as the Nevada test site in the American Southwest, Maralinga test site in South Australia, Bikini and Mururoa Atolls, etc., all surface detonations

much more akin to Chernobyl or Fukushima. You don't want to live there! Japan was able to rebuild Hiroshima and Nagasaki because the bombs detonated above the cities and didn't produce neutrons.

Even getting the uranium to supply our reactors is hazardous. Uranium mining releases radon and other pollutants into the atmosphere and leaches them into the ground polluting streams, springs, and our drinking water.

As luck would have it, a large deposit of uranium ores exists on the Navajo Nation here in the Southwest. Beginning in the 1940's, widespread mining and milling of this ore for national defense and later for nuclear energy, led to a plethora of uranium mines leaching radiation and heavy metal contamination into the air, soil and water. Old waste piles rich in radioactivity dry out in the sun allowing the wind to pick it up and spread it across vast areas. Many Navajos have fallen sick with anemia, cataracts, and bone cancer, far higher than the general population. On November 4, 1993, the Navajo Nation testified to a Congressional hearing along with the EPA[7], DOE[8] and BIA[9] about the problem. The EPA offered to assist under Superfund Law and initiated the first of many studies in 1994 aimed at assessing human exposure to radiation and heavy metals from every known abandoned uranium mine (AUM) on the Navajo Nation.

It wasn't until August 2007, after yet another larger study that the EPA identified 520 individual AUMs. The worst is located near Gallup, New Mexico. North East Church Rock is the highest priority abandoned mine cleanup on the Navajo Nation. The mine adjoins an old United Nuclear Corporation uranium mill, a Superfund site (HR 30.36) managed jointly by EPA and the Nuclear

7 Environmental Protection Agency
8 Department of Energy
9 Bureau of Indian Affairs

Regulatory Commission. The mine is on tribal land and not listed as a Superfund site, but at the request of the Navajo Nation, the EPA is using Superfund authority to investigate and clean up the mine site along with the mill site.

Since then, they removed 6,500 cubic yards of uranium-contaminated soils from the yards and property of four Gallup residences costing taxpayers $2,200,000. At another nearby community, Red Water Pond, the EPA removed over 100,000 cubic yards of uranium-contaminated soil from the yards, arroyos, and the surface of a local road but their groundwater remains contaminated. How do you decontaminate an aquifer? The EPA advised local residents not to drink from it but many do. After all, they have been drinking from these same wells for a thousand years. Old habits die hard.

In 2011, the EPA announced a new effort to clean up the Northeast Church Rock Mine. They plan to remove another 1.4 million tons of uranium-contaminated soil, placing it in a nearby lined and capped facility,[10] all at taxpayer expense.

All this activity and money spent on cleaning up old uranium mines seems to have escaped Arizona State Senator Sylvia Allen (R-Snowflake). In July 2009, she argued in favor of a bill allowing uranium mining north of the Grand Canyon, casually saying twice:

..this Earth's been here 6,000 years, long before anybody had environmental laws and somehow hasn't been done away with (yet). We need to get the uranium here in Arizona so this state can get the money for it.

Arizona State Senator Sylvia Allen

Apparently, a little uranium mining isn't going to hurt anything as long as someone is making money. Snowflake is a scant 100 miles southwest of Gallup, New Mexico, where the EPA is cleaning up the worst of 520 abandoned uranium mines on the Navaho Nation, many of them in Arizona. Snowflake is just south of the reservation and part of Navaho County, Arizona. Perhaps Senator Allen should take a short drive and ask a few Native Americans the effects of uranium mining on them. The blatant disregard for facts is enough to make your head explode.

Senator Allen should know that when we dig up uranium ore and process it, the waste this generates still contains a much higher level of residual uranium than normal soil resulting in a marked increase in radon levels. Mining brings uranium to the surface where it can do the most damage.

Radon (atomic number 86) is a nasty little element and the heaviest naturally occurring noble gas in the Periodic Table.[11] It is odorless, colorless, and tasteless, impossible to detect using just human senses. It forms as part of the normal decay chain of uranium and thorium. The most common isotope of uranium has a half-life of 4.5 billion years, and thorium at just over 4 billion. This fact is one of the main ways we measure the lifespan of the Earth. However, the most common radon isotope

Rn 86 (222)

Density
9.73 g/L

Boiling point
-62°C

F.E. Dorn, 1900

Melting point
-71°C

California Geological Survey Mineral Resources and Mineral Hazards Mapping Program

$(Xe)\ 4f^{14}\ 5d^{10}\ 6s^2\ 6p^6$

Radon

---

10 Superfund Site, EPA

11 Dayah, Michael. Dynamic Periodic Table. 1 October 1997. 4 January 2012

has a half-life of just 3.8 days, which makes it seem harmless at first.

The problem is that as the radioactive radon gas decays, it produces new radioactive elements called radon daughters. Radon daughters are solids and stick to surfaces such as dust particles in the air. When we inhale contaminated dust, these particles can lodge in the air passages inside your lung where they will cause cancer. *It just takes one!*

The EPA estimates that radon causes about 21,000 cancer deaths in America each year.[12] Radon is the complete carcinogen because, unlike chemicals, a single radon atom can initiate, promote and propagate lung cancer. Children are particularly susceptible. Mine waste from operations that closed before the mid-1970s are of particular concern. Weathering leads to radioactive dust picked up by the wind and spread across public and private lands. Not only will it spread far and wide through the air, the relentless seepage of toxic contaminants into the ground will eventually reach the local aquifer or even find its way into your home through the walls of your basement.

Nuclear energy is dirty at both ends. A typical power plant generates twenty metric tons of spent fuel in a year. All together, the industry generates about 2,200 metric tons per year. Spent fuel rod assemblies, each containing roughly 380 pounds of uranium, have been accumulating since December 2, 1942, when the Chicago Pile-1 became the world's first nuclear reactor. It was part of the Manhattan Project during World War II. The Italian physicist Enrico Fermi supervised the project. Since then, the entire industry has produced about 65,200 metric tons of used nuclear fuel, all of it stored in pools within the facilities that generated it. The graphic lists the tonnage by state.

Figures provided by Tokyo Electric Power show that most of the at risk uranium at the stricken power plant is actually in the spent fuel rod assemblies, not the reactor cores themselves. Fukushima is showing the world how dangerous it is to store spent fuel rods on site, but with no

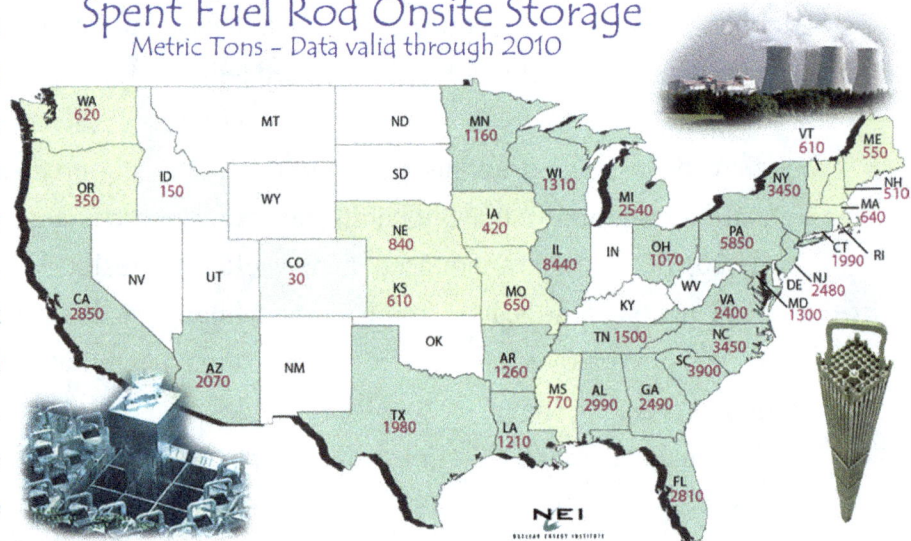

Spent Fuel Rod Onsite Storage
Metric Tons - Data valid through 2010

NEI

long-term storage, the nuclear industry has no choice.[13]

The only safe location for such dangerous material would be one that is both geologically stable and isolated from all contact for the next 1,000,000 years. Sound impossible? It is. About 80 miles (130 km) northwest of Las Vegas, Nevada, Yucca Mountain Nuclear Waste Repository was the planned storehouse for our nations spent nuclear reactor fuel and high-level radioactive waste.[14] The Obama administration

12 Dangers of Radon, EPA

13 3/17/2011: Greater Danger in Spent Fuel Than in Reactors, NYT, K. Bradsher
14 12/19/2011: Search is on for nuke-waste site, Wire Service, Arizona Republic

canceled the project in 2009. Seems our fellow citizens in Nevada didn't want the repository in their backyard or using their roads. The Department of Energy is reviewing other options but for now, America is without any long-term storage solution.

One answer is to reprocess the spent fuel and use the recovered fissile material to power our reactors some more. However, even though the idea sounds good, there is great danger in doing it. Keep in mind that nuclear power has its roots in the cold-war bomb makers. The same facilities that enrich uranium for use in power plant fuel can make highly enriched uranium (HEU) used in bombs while facilities that reprocess spent reactor fuel produce plutonium. Nations, corporations, or individuals possessing those facilities can build nuclear weapons.[15]

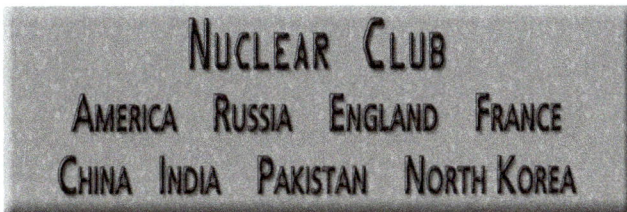

NUCLEAR CLUB
AMERICA   RUSSIA   ENGLAND   FRANCE
CHINA   INDIA   PAKISTAN   NORTH KOREA

Thirty countries operate about 440 nuclear power plants. Any expansion of nuclear energy will certainly increase the number of unfriendly nations that have weapons grade material and thus, access to nuclear weapons. America became the founding member of the Nuclear Club when we detonated the world's first A-bomb on July 16, 1945. The first Soviet test followed four years later on August 29, 1949, Great Briton on October 3, 1952, France on February 13, 1960, and China on October 16, 1964. India's first nuclear weapons test occurred ten years later on May 18, 1974. On May 28, 1998, Pakistan detonated five weapons in their first test. North Korea is the latest to join the club detonating their first nuclear weapon on October 6, 2006. In January 2012, Iran announced they have produced their first nuclear-fuel rod marking a major milestone in their quest to join the club. [16]

One thing the United States could do immediately is reinstate the ban on reprocessing spent fuel and actively discourage other nations from pursuing reprocessing. I personally don't think this will happen. Countries like Iran and North Korea will not abide by our wishes, not to mention China. The power associated with enrichment and reprocessing technologies is too great. Nuclear proliferation will continue regardless of what policies we pursue here in America. The only way to slow this is to offer something better. **SBSP** for instance.

Even if we accept the dangers inherent in nuclear energy, how many power plants would we need to build? Let's start by looking at what currently exists. America has the most with a 104 nuclear power plants accounting for about 20% of our electrical needs. To power our country with uranium would require a total of 520 plants or 10 in every state including Rhode Island. Perhaps we should add in a few more to cover the increase in electrical demand while we build all these multibillion-dollar monstrosities.

Before the Fukushima Disaster, there were 440 operating nuclear plants around the world with 63 under construction. The top spots on the list starts with America at 104, France is next with 58, Japan 54, Russia 32. China is coming on strong with 13 operational and 27 under construction. The percentage of electricity supplied by nuclear ranges from a high of 78% in France, 54% in Belgium, 39% in Republic of

---

15 Preventing Nuclear Proliferation and Nuclear Terrorism, Union of Con Sci

16 1/2/2012: Iran says it has produced its first nuclear-fuel rod, Ali Dareini, AP

Fukushima Daiichi Power Plant

Korea, 37% in Switzerland, 30% in Japan, 20% in the USA, 16% in Russia, 4% in South Africa, and 2% in China. Just to meet the current demand for electricity using nuclear energy, the world would need to build over 21,500 new plants. That's just not going to take place.

Especially given what just happened to Japan, the world is having nuclear second thoughts including China. The Associated Press reports that China suspended approvals for new nuclear power plants and is conducting additional safety checks at existing plants and those under construction. Their plan was to reduce reliance on coal-fired plants by leaning heavily on nuclear power. We'll see if that happens. China also leads the world in developing solar, wind and hydropower. Ramping up that endeavor may be the better choice but more likely, they will turn to coal or natural gas.

Each reactor requires an average of 135 tons of uranium to begin operation, which means we will need over 3 million additional tons of this dangerous metal! Even if you can ignore this beast living alongside you, just envision the number of semi trucks and rail cars it will take to move the massive amount of uranium around to support such a monster. Do you really want to share your world with so much uranium? The occasional fuel tanker blowing up on the interstate or BP crude oil spill off our coast is bad enough, but radiation contamination is a billion times worse. Do you really want to put so much highly toxic and unimaginably dangerous material where terrorists can get their hands on it? It is not a question of *if* an accident or incident will occur, but which one happens first.

Yet, even when the dangers of nuclear energy are so clearly demonstrated, there are people who will tell you it is perfectly safe. While the Fukushima disaster was still playing out on the world's high definition flat screens, CNN put on Jay Lehr, a self-proclaimed expert in nuclear power and Science Director of the Heartland Institute. Mr. Lehr arrogantly assured us that nuclear energy is safe even as the disaster was occurring. He then smugly informed everyone, *"no matter the dangers, nuclear energy is the power source of the future. Get used to it."* I emphatically disagree. **Space Based Solar Power** is the way to go.

What CNN or Jay didn't inform you was who finances his opinion. The 2007 list of energy companies that contribute to the Heartland Institute reads like a who's who in the industry. ExxonMobil, Chevron Corporation, ConocoPhillips and specifically, Koch Foundations.[17] Charles Koch supports Jay Lehr and the Heartland Institute with his money. I'm

---

17 Heartland Institute - Koch Industries Climate Denial Front Group, Greenpeace

sure Jay is well paid to express the right opinion. I'm using 2007 data because the recent change in the law means the institute is no longer required to list its major donors. That's what we need, more secrecy.

Thorium-232 is a radioactive metal proposed as an alternative to uranium fuel in nuclear reactors. In the old days, like uranium, thorium was found using Geiger counters. Today, satellite gamma-ray spectrometry maps out vast deposits around the world. Thorium is about three or four times more abundant than uranium in the natural environment. America has plenty and there are even deposits on the Moon if we run out down here on Earth. Thorium proponents claim that it solves the proliferation, waste, safety, and cost problems of uranium-fueled nuclear power. Many very knowledgeable people disagree.

The biggest problem is that thorium is not fissile and therefore, cannot start or sustain a nuclear chain reaction. Thorium reactors, when and if they come online, will need uranium or plutonium to kick-start the reaction. Granted the system would not use as much as before but these highly dangerous materials are still part of the process.

The thorium-232 in the spent fuel has a half-life of 14 billion years and its decay products will build up over time making the spent fuel quite radiotoxic. The inhalation of thorium-232 or thorium-228 produces far more damage than the equivalent quantity of uranium. The damage from breathing thorium is about 200 times that of inhaling the same amount of uranium.[18] Nice stuff.

The fact is, it takes a PhD and years of study to understand the complexities of nuclear physics and I sometimes wonder if anybody knows what's really going on, but one thing is clear; switching to thorium is not the answer. Thorium fuel does have some advantages but for the most part, it would be exchanging one bad answer for a different bad answer to our energy future. Nuclear energy will have its place off planet, powering our ships and helping establish communities in very inhospitable places as we push humanity's frontiers to the Moon and beyond but back home on planet Earth, it is just too dangerous. **SBSP** is a much safer solution.

Let me emphasize again, radiation contamination is not a simple cleanup like an oil spill or a burst pipeline or even the mind-numbing devastation of the Gulf War Oil Disaster. Radiation contamination means a total abandonment of a significant area of land for centuries or millennia. On a human time scale, this might as well be forever. With so much radioactive material around, used and unused, processed and reprocessed, how long before some crazy gets their hands on some? You don't need high-quality uranium or weapons grade plutonium to wreak havoc on the world. Using an Improvised Explosive Device (IED) packaged with a few pounds of radioactive waste will become the poor man's nuclear weapon.

The proliferation of nuclear power guarantees cheap dirty bombs, big powerful ones and occasionally, natural disasters spreading radiation across the land. We should avoid such a devastating future at all costs and the only way to do that is to abandon nuclear energy now, before we invest more money on it, before we build a system that guarantees disaster. There must be another way to generate energy other than letting Pandora out of her Box. There is: **SBSP**.

18   Thorium Fuel: No Panacea for Nuclear Power, By Arjun Makhijani and Michele Boyd, A Fact Sheet Produced by the Institute for Energy and Environmental Research and Physicians for Social Responsibility

Water and Electricity

The various ways we generate electricity requires lots of water.[1] Hydroelectric dams return most of the water they use to generate electricity to the source and use up only about 70 gallons for every 1000 kilowatt-hours.[2] Geothermal (450 gal) and solar thermal (850 gal) are also fairly water efficient. Fossil fuels, however, are not. Fossil-fuel-fired power plants consume more than 130 billion gallons of fresh water per day in the United States alone.

Coal mines use millions of gallons of water just to wash the coal dust from their machines. A typical deep shale gas or oil well uses 4.5 million gallons of water to drill and fracture the subsurface rock. Water keeps power plants cool, removes pollutants from power plant exhaust, flushes away residue after fossil fuels are burned and, last but not least, steam turns the turbines that generate electricity. Yet, coal and gas form the backbone of electricity production worldwide using up our planets water resources even as they pollute our world.

Of the major thermoelectric power generating methods, nuclear power is the least efficient. Palo Verde, the 3-gigawatt power plant just down the road from me, uses 20,000,000,000 gallons each year to bring me electricity.[3] I live on a desert but if there's one thing the Fukushima Disaster has shown us, we don't want our nuclear power plants to run out of water. If the core doesn't get you, the spent fuel storage pool will. Our existing nuclear plants will need water for at least fifteen years after we shut them down and any new ones will

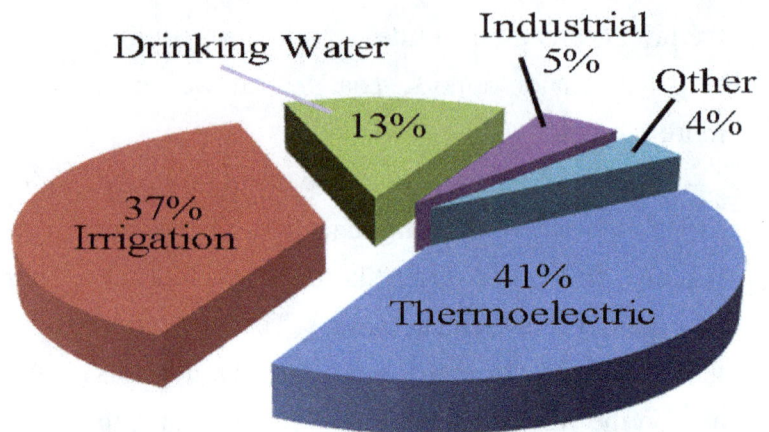

U.S. Freshwater Withdrawals
Power plants account for the largest share of freshwater usage in America

add to the debt. When we add up the total usage, data from the U.S. Geological Service shows we are using over 200,000,000,000 gallons of water every day to produce our electricity.[4]

Very briefly, synthetic fuels use hydrocarbon feedstock such as coal or natural gas, which use water to mine and refine. Corn and soy-based biofuels, biodiesel, and ethanol production use a lot of water, crossing over into the agricultural realm where irrigated crop production is the next biggest consumer of water.

All over the world, water is becoming an issue of life and death. On an isolated atoll in the South Pacific Ocean, 10,000 people wonder where their next drink of water will come from. They can no longer count on the fresh water wells that have supplied the nation of Paleli Tovia for centuries.[5] Rising sea levels and urban pollution have contaminated them beyond use. In another ominous indication of coming water shortage, the snowpack along the western face of the Rockies is about 75 percent of average, bad news for the states downstream along the Colorado River,

1 The connections between our energy and water, Union of Concerned Scientists
2 Energy Technologies, Union of Concerned Scientists
3 Palo Verde, APS
4 Water use in the United States, National Atlas
5 10/16/2011: Island Nation's water crisis may be a warning for world, Wire Service, Arizona Republic

GRACE-based Surface Soil Moisture
January 06, 2014

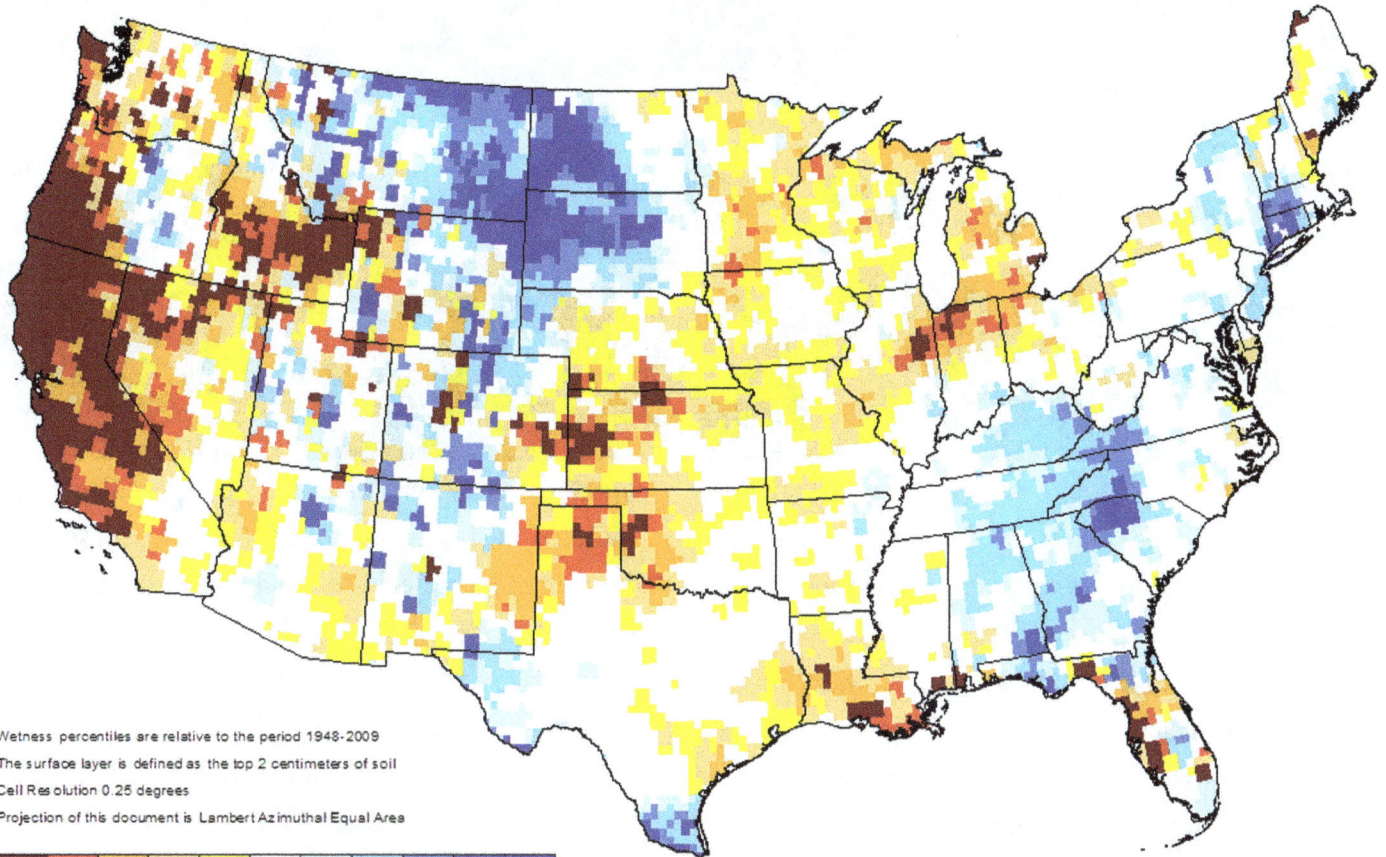

Wetness percentiles are relative to the period 1948-2009

The surface layer is defined as the top 2 centimeters of soil

Cell Resolution 0.25 degrees

Projection of this document is Lambert Azimuthal Equal Area

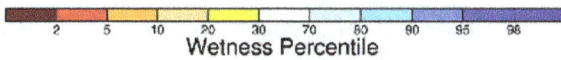

Wetness Percentile

2 5 10 20 30 70 80 90 95 98

http://drought.unl.edu/MonitoringTools/NASAGRACEDataAssimilation.aspx

Arizona, Nevada, Utah and California[6] where there is a growing drought of biblical proportions.

Mexico is looking to make a buck selling water to these states. They are planning to build two huge energy-intensive desalination plants along the Mexican coastline just south of San Diego capable of supplying 200 million gallons a day.[7] Why not invest that money into **SBSP**, eliminate the high usage power plants and reallocate their water to other needs?

Although some technologies are more water efficient than others, water availability will only decrease as our energy demands increase.

Worldwide population growth and increasing energy consumption will continue to put pressure on our water supply and compete with food production. It is a viscous cycle. Are you willing to trade a drink of water for a kitchen light? There is a critical need for a type of energy production that uses less water, not more.

There is only one power source capable of taking the place of coal, gas, and nuclear and that's the Sun. **Space Based Solar Power** requires very little water during assembly and virtually none during operation.

6 1/12/2012: Colorado snow pack below average, Wire Service,
   Arizona Republic
7 10/16/2011: Mexico looks to export water to Western states, AP

We Reap What We Sow

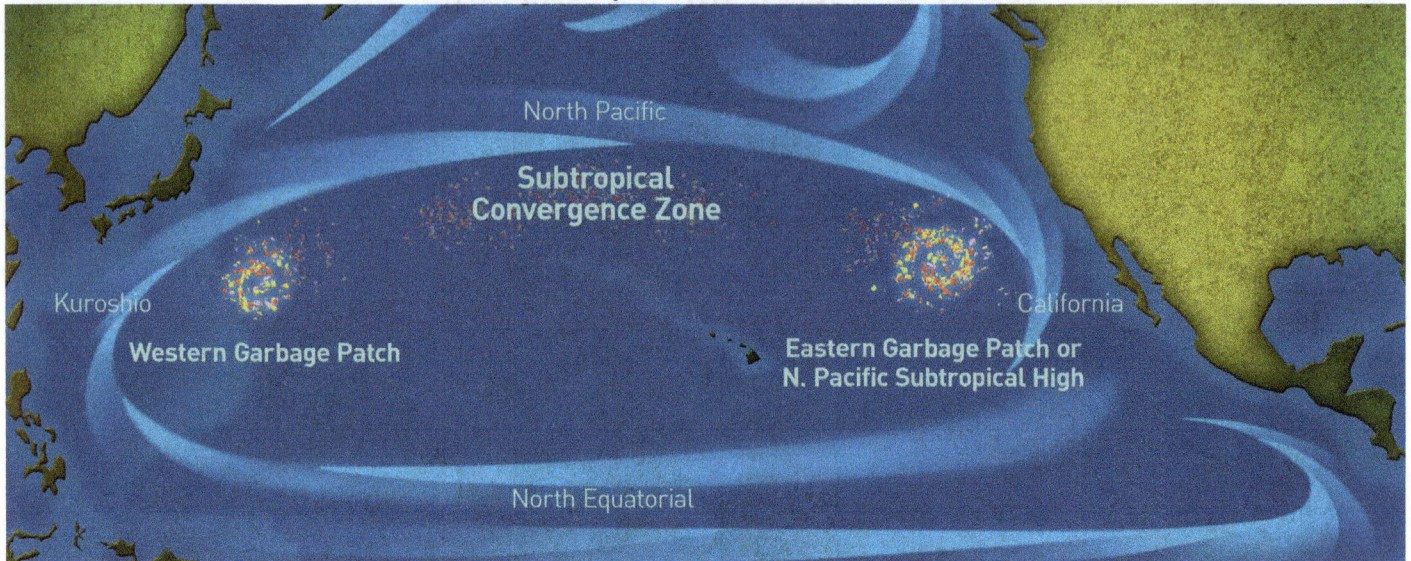

North Pacific

Subtropical
Convergence Zone

Kuroshio

California

Western Garbage Patch

Eastern Garbage Patch or
N. Pacific Subtropical High

North Equatorial

Are you old enough to remember when going to the beach was reason to celebrate? Those days are over. Every North American coastline including the Great Lakes is contaminated. High levels of bacteria, viruses and parasites in the water resulted in 20,000 beach-closing and swimming advisory days just in the USA alone.[1] For the sixth year in a row, raw sewage and storm runoff gathered in an ominous plume offshore along California beaches creating an environment so bad that people became ill by swimming in them for just a few minutes.[2] Throughout the summer of 2009, five Washington state beaches closed repeatedly due to Methicillin-resistant Staphylococcus aureus (MRSA).[3]

In August 2009, an expedition from Scripps Institute of Oceanography at the University of California, San Diego, set sail for the Great Pacific Ocean Garbage Patch. They were gone for less than a month but what they learned is astonishing. About 1000 miles west of San Diego, global ocean currents create an eddy that traps our garbage.[4] It is roughly the size of Texas and filled with countless bottles, Styrofoam, tangled nets and old patio chairs. You know the ones I'm talking about, the cheap plastic ones down at Wal Mart. This is where plastic bags go to die.

As bad as that is, the story doesn't end there. Most alarming was the nearly inconceivable amount of tiny confetti-like bits of shredded plastic they found in water samples.[5] Not just in the immediate vicinity of the Great Patch, but permeating the entire ocean. They collected hundreds of water samples, crossing thousands of miles of open ocean and found plastic confetti in *every... single... sample*, one-hundred percent, without exception!

Plastic is not a natural substance. It does not occur in nature. Plastic is strictly a man-made polymer composed of ethylene, propylene, vinyl, or styrene, and processed with a number of toxic compounds. No one knows the long-term affect these complex hydrocarbons will have on fish and other ocean life but one thing is certain, whatever happens, we're responsible.

1 7/29/2009: Raw sewage, other waste still plague U.S. beaches, Wire Service, Arizona Republic
2 7/29/2009: Calif. Beach contamination increasing, study shows, Wire Service, Arizona Republic
3 9-13-2009: Resilient staph bacteria found on Washington (state) beaches, Wire Service, Arizona Republic

4 8/5/2009: Expedition to study patch of plastic garbage in ocean, Wire Service, Arizona Republic
5 9/4/2009: Garbage expedition pulls piles of debris from ocean, Wire Service, Arizona Republic

For many years, we hauled our trash out to sea and dumped it. In 1975, the United Nations implemented a worldwide ban on ocean dumping but it wasn't until 1988 that our government passed the Ocean Dumping Ban Act putting an end to it in America. In 1996, the EPA declared 17 square miles of ocean off the Palos Verdes Peninsula, California, a Superfund site due to our using the ocean as a landfill. Lying 200 feet underwater, it is the world's largest deposit of banned pesticide DDT. Montrose Chemical Corporation released 110 tons of DDT and 10 tons of toxic PCBs into the sewers from 1947 through 1971 where it then flowed into the Pacific and collected in the mud and sand just offshore.[6]

Plastics are not the only way complex hydrocarbons are getting into our environment. We move hundreds of thousands of tons of hydrocarbons in pipes, ships, and trucks, all over the world. Accidents happen and people make mistakes. It is inevitable the oil and other toxins will get into our environment.

Let's start with the most celebrated oil spill in American history. You know the one I'm talking about, where the drunken captain let his ship run aground in Prince William Sound, Alaska. On March 24, 1989, the tanker *Exxon Valdez*, was bound for Long Beach, California when it struck Bligh Reef.

Within hours of the grounding, the *Exxon Valdez* spilled 10.9 million gallons of its 53 million gallon cargo of crude oil. The oil would eventually affect over 1,300 miles of Alaskan coastline and cover 11,000 square miles of ocean, making this the largest oil spill in American waters… at least until the BP Gulf Disaster of 2010. Will it surprise you to learn that worldwide, the *Exxon Valdez* spill ranks in the mid thirties in number of gallons released? It did me.

The list of oil spills is very long and should be

a lot longer, but where do you draw the line? Minor spills happen all the time. What is your definition of a spill? Anything above a thousand gallons? Ten thousand? Most of these go unreported. GPS and double-hulled supertankers have slowed down the frequency of major spills but not eliminated them. What follows is a list of history's top ten crude oil spills. Careful your fingers don't get oily.

I. **1991 – Gulf War Oil Disaster, Kuwait - 75 billion gallons:** Billion is not a typo. Nearly 1,800,000,000 barrels (42 gal per barrel) of oil burned in the Kuwaiti oil fires. 732 wells were set afire, while many others were severely damaged and gushed uncontrolled for months. The fires alone consumed approximately 6,000,000 barrels of oil per day at their peak. The oil that didn't burn pooled in hundreds of lakes and absorbed into the ground forming a layer known as tarcrete extending over approximately five percent of the surface of Kuwait. In a separate disaster, Iraqi forces opened the valves of several oil tankers in Kuwait Harbor to slow the invasion of American troops. The oil slick covered an area of ocean larger than the island of Hawaii and 5 inches thick in spots. What happened in 1991 is actually hundreds of disasters rolled into one. Together, Kuwait is the largest spill in our history, so far.

II. **2010 - BP Gulf Disaster, Gulf of Mexico - 206 million gallons:** April 20, 2010, an exploratory oil well in the Gulf of Mexico exploded at the wellhead 5000 feet below the surface of the water. Unable to deal with the situation, the well remained open until 15 July 2010 spewing 2,600,000 gallons a day into the ocean at its height. If not for the total disaster in Kuwait, this oil spill would rank as the largest in history… The location of the spill, a mere 41 miles off the Louisiana coast and the Mississippi delta, affected some of the richest marine ecosystems in North America.

III. **1979 - Ixtoc 1 Oil Well, Gulf of Mexico - 126 million gallons:** June 3, 1979, an exploratory oil well in the Gulf of Mexico exploded at the wellhead 160 feet below the surface of the water. Unable to deal with the situation, the well remained open until 23 March 1980 spewing 500,000 gallons a day into the ocean at its height. For a short twelve years, this oil spill

6   6/12/2009: EPA seeks to clean up DDT tainted site off Palos Verdes
    Peninsula, Jeff Gottlieb, LA Times

reigned as the largest in history.

IV. **1979 - Atlantic Empress, Caribbean Sea - 90 million gallons:** During a raging storm on the Caribbean Sea off the coast of Tobago, two full supertankers collided. Crippled and leaking crude oil, both caught fire. The fire on the **Aegean Captain** was soon controlled, but the other tanker, the **Atlantic Empress,** exploded killing 26 crewmen and releasing its entire load of crude into the ocean.

V. **1992 - Fergana Valley Spill, Uzbekistan - 88 million gallons:** The Fergana Valley Oil Spill started on March 2, 1992 at the Mingbulak oil field when a blowout occurred at a major wellhead. The oil caught fire and burned out of control for two months. They made dykes to hold what didn't burn recovering about 80 million gallons but they were never able to plug the hole themselves. The oil stopped flowing all by itself. The spill didn't get much press at the time even though it was the worst oil spill on land in history. Very few even noticed.

VI. **1983 – Nowruz Spill, Persian Gulf - 80 million gallons:** Two platforms in the Nowruz oil field spilled over 80 million gallons of crude into the Persian Gulf. The year started badly when a tanker hit a drilling platform damaging the well and starting it leaking. During repairs, Iraqi helicopters attacked the platform and set the spill on fire. Helicopters also attacked a nearby drilling platform causing it to start leaking. The Iran-Iraq War prevented technicians from capping the first well until September 1983, the second not until May 1985.

VII. **1991 - ABT Summer, Atlantic Ocean - 80 million gallons:** 700 miles off the Angolan coast, the fully-loaded supertanker **ABT Summer** began leaking oil and caught fire. It burned for three days before sinking. It leaked its entire cargo into the ocean. The oil slick covered 80 square miles.

VIII. **1983 - Castillo de Bellver, Atlantic Ocean - 79 million gallons:** On August 6, 1983, the supertanker **Castillo de Bellver** caught fire about 70 miles off the coast of South Africa. They had to abandon ship and the blazing tanker drifted offshore until it eventually broke apart and sank releasing its cargo some 25 miles off the coast of Saldanha Bay.

IX. **1978 - Amoco Cadiz, English Channel - 69 million gallons:** The steering rudder of the tanker **Amoco Cadiz** failed in a severe storm, despite the efforts of several ships, the ship ran aground off the coast of Brittany and broke in two. Its entire cargo of crude oil spilled into the English Channel.

X. **1991 - M/T Haven, Mediterranean - 42 million gallons:** The dilapidated tanker, **M/T Haven**, caught fire and exploded off the Italian coast killing six and later sinking in flames. It continued leaking its oil into the ocean for 12 years. At 820 feet long, **M/T Haven** remains one of the largest shipwrecks in the world and is a popular tourist destination for divers.

With no new major oil fields discovered, energy companies have begun drilling in some very precarious spots. The recent BP Gulf Disaster is a warning that we must heed. We cannot put the fate of our world into the hands of companies whose major motivation is profit. Corporations are set up to make money, not friends. We should know by now that the assurances given by oil companies mean little before or after an accident happens. In the Gulf, it quickly became obvious that BP didn't know how to deal with a problem it had created. Keep in mind the Gulf Disaster happened off our southern coast, right under our noses. The problems will be much worse when and if we drill in the Arctic and Antarctic. What do you think will happen when it is thousands of miles away and in sub-zero weather? Out of sight, out of mind? Will the average citizen even take notice? Can you blame them? For most of us, life's daily routine takes precedence. We must rely on our government to do the right thing and not be in the industry's pocket.

For every major oil spill, there are thousands of minor ones. For example, in late February 2010, saboteurs released 660,000 gallons of oil sludge into the Lambro River near Monza Italy.[7] It quickly reached the Po River, Italy's longest, where it contaminated irrigation water for a large swath of the country.[8]

On November 7, 2007, a cargo ship sideswiped a tower on the San Francisco-Oakland Bay Bridge and spilled 53,000 gallons of oil contaminating 26 miles of California shoreline.[9]

In July 2011, an ExxonMobil pipeline in Montana ruptured, leaking crude oil into the Yellowstone River. The company first claimed that only a few hundred gallons escaped but had to quickly revise the total into the tens of thousands. This list is endless…

We should always keep in mind when choosing our path that pipes leak and accidents happen. We hope for the best but always prepare for the worst. People will cut corners and make bad decisions selfishly seeking profit or power and there's nothing we can do to stop it. It is human nature. What we can do is set up regulatory agencies to oversee the most dangerous activities, those things effecting people's lives and livelihoods.

Another consequence of spewing massive amounts of CO_2 into the atmosphere is the acidification of water. One of our planet's natural mechanisms is the ability of our oceans to absorb carbon dioxide. In fact, our oceans are capable of absorbing roughly one-third of the gas released by human activity. The problem is that when CO_2 dissolves in water, it forms carbonic acid that decimates marine life. Carbonic acid dissolves calcium carbonate in seashells and coral reefs destroying them.[10] Fish larvae are particularly susceptible to carbonic acid, killing off 70% within a week due to organ damage. The fish that do survive to maturity ends up much smaller than normal.

Not only are we contaminating the oceans at a horrendous rate, rising temperatures are melting the polar ice caps as well, contributing to a substantial rise in sea levels around the world. Scientific measurements show the oceans are rising 1/10 of an inch per year, an increase of 550% above what it was just twenty years ago.[11] At that rate of increase, it will be rising an inch per year by 2030 and a foot per year by 2050. By the end of this century, rising sea levels will total 65 feet, by the end of the 22nd century over 200 feet.

Rising ocean temperature also makes the water expand contributing to rising sea levels. If all the ice melts, and global temperatures continue to rise, sea levels will exceed 260 feet from what they are today. This will radically alter our shorelines and drive billions of people from their ancestral homes. Like mythical Atlantis, islands, coastal cities, and low-lying countries will sink below the waves. Bays and inlets will reshape themselves and what once were fertile farmlands will become shallow inland seas.

The polar caps aren't the only melting ice. Tens of thousands of glaciers on the Tibetan Plateau form the headwaters for seven major Asian rivers. The Himalayan range contains over 15,000 glaciers storing about 12,000 cubic kilometers (km^3) of freshwater. With nearly 37,000 glaciers on the Chinese side alone, the Tibetan Plateau and its surrounding arc of mountains contain the largest volume of ice

7 2/25/2010: Oil spill threatens Italy with Ecological disaster, Wire Service, Arizona Republic
8 2/26/2010: Oil slick reaches Parma key farming area of Italy, Wire Service, Arizona Republic
9 2/20/2010: Cargo-ship operator fined $10 million in Bay oil spill, Wire Service, Arizona Republic
10 2/5/2009: Ocean Acidification from CO2 is Happening Faster than Thought

11 4/14/2011: Coastal Zones and Sea Level Rise, Environmental Protection Agency

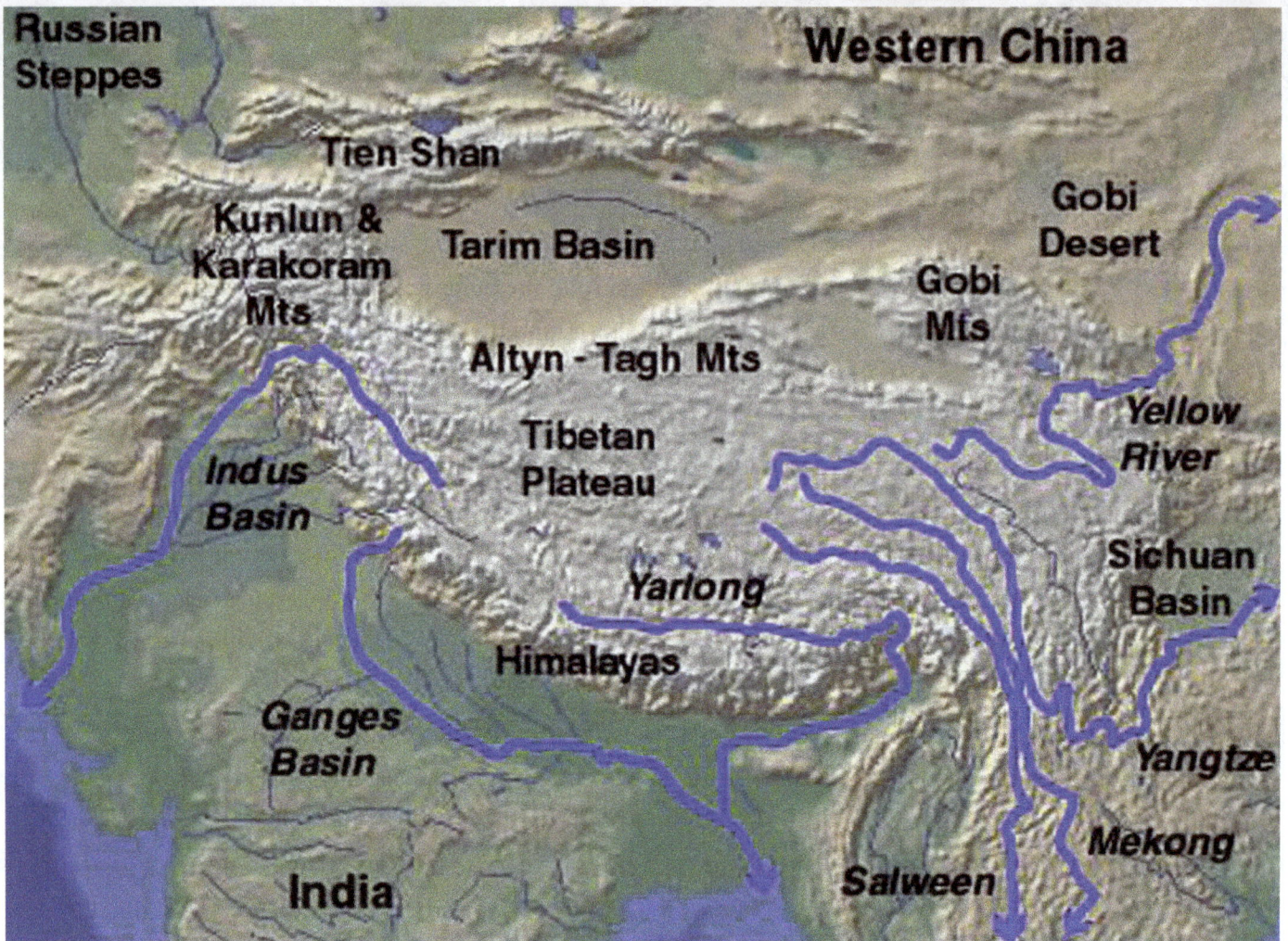

outside the Polar Regions. At 70 km, the Siachen Glacier at the India-Pakistan border is the second longest glacier in the world. Collectively, the Tibetan glaciers generate the largest run-off from any single location in the world, water for three billion people, almost half the world's population.

According to the United Nations report, *Global Outlook for Ice and Snow*, virtually all of these critical glaciers have retreated during the 20th century but this shrinkage has accelerated substantially during the past decade. That's bad news for those people who live downstream. The increased glacier run off mixing with spring rains cause severe flooding. Glacier melt can pose an even greater threat to the river valleys when ice dams accumulate vast lakes behind them. Outbursts send a wall of water downstream with the power to breach man-made dams and overwhelm communities.

What happens if the glaciers completely disappear? The World Wildlife Fund lists the Indus as the largest river in the world at risk of dying because of climate change. The Indus begins its 2,900 km journey to the sea in Tibet at a spring known as *The Mouth of the Lion* and flows through India, China, Pakistan and Afghanistan. It is the lifeblood of almost a half billion people and already suffers from severe water scarcity due to over extraction for agriculture. With the demise of the glaciers, the extended outlook is grim, to say the least.

Superfund and You

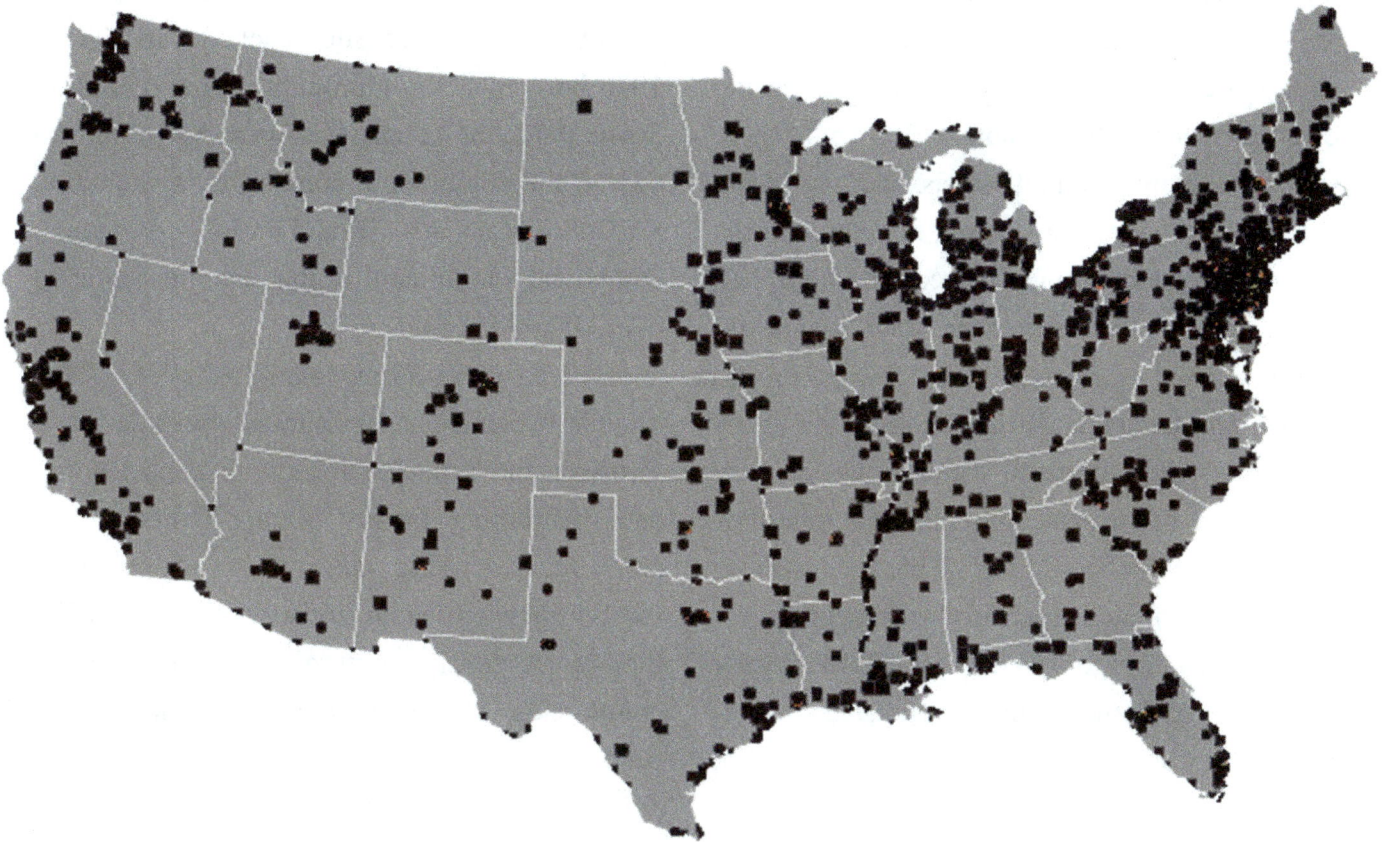

Not only are we polluting the world's oceans, we are contaminating the ground we live on at a record pace.

In the middle of the last century, following two world wars, America began writing laws designed to protect citizens and the environment from industrial chemicals. One of the most radical of them provides a means of cleaning up contaminated sites. In 1980, President Carter signed into law the ***Comprehensive Environmental Response, Compensation, and Liability Act*** (CERCLA), commonly referred to as the Superfund. This legislation also instructed the Environmental Protection Agency (EPA) to identify parties responsible for contamination of sites and compel those parties to clean them up. Where responsible parties no longer exist, known as orphaned sites, the law mandates the EPA to

clean them up. That's you and me, taxpayers cleaning up another corporate mess. Originally, taxing the energy and chemical industries they regulate funded EPA efforts, but the last year the IRS collected those taxes was FY1995. The Republican *Contract With America* stopped that practice. Congress must appropriate the money on a case-by-case basis to clean up an orphaned Superfund site out of general revenues. Now there's no question about it, taxpayers are footing the bill, if it gets cleaned up or not. More and more, it is just not happening. Studies have taken the place of action.

Here's how it works. A potential site is evaluated, given a priority and placed on a list. The EPA uses a complicated rating called the Hazard Ranking System (HRS) to determine where the potential site should be placed on the National Priorities List (NPL). They have a training course to teach you its intricacies and an online questionnaire that determines your sites grade. The final score combines the specific characteristics of the contamination with where it is, how many people it affects, and how many ways does it have to spread out to the rest of the world. The scores range between the high 20's to the mid 70's, the higher the score, the nastier the site.

As of January 1, 2012, America had 1,298 Superfund sites listed on the National Priority List, with an additional 354 delisted and 62 new sites under evaluation. Each of these constitutes an environmental disaster of various degrees.

(HRS 74.86) From 1942 to 1990, McCormick & Baxter Creosoting Company operated a wood-preserving business in Stockton, California, where they treated utility poles and railroad ties with creosote, pentachlorophenol (PCP), and arsenic compounds. The 29-acre site is in a light industrial area near the Port of Stockton. Old Mormon Slough, a tributary to the San Joaquin River, borders the site to the north. Waste generated from the wood-treatment processes were disposed of in unlined ponds and concrete tanks on-site. Surface water runoff from the site discharged to the slough until 1978, when the company installed two collection ponds. In 1984, studies found that soils throughout the site were contaminated with arsenic, chromium, copper, PCB, and polycyclic aromatic hydrocarbons (PAHs), all constituents of creosote. Soil contamination extends downward to a shallow aquifer interconnected with the regional deep aquifer that provides drinking water to approximately 97,000 people.

(HRS 69.92) New Castle County in Delaware owned and operated the Army Creek Landfill, a municipal and hazardous waste disposal facility. The 47-acre site reached capacity in 1970 holding 1.9 million cubic yards of refuse. Since 1972, the county has spent $3 million to control the migration of contaminants, including lead, chromium, arsenic, and a variety of organic compounds, to an aquifer that supplies water to over 100,000 people.

(HRS 70.71) From 1893 to 1978, Kerr-McGee Chemical Corporation operated as a fertilizer and pesticide formulating, packaging, and distributing facility on approximately 31 acres located along the western shoreline of the St. Johns River in downtown Jacksonville, Florida. A deadly mixture of volatile organic compounds, semi-volatile organic compounds, pesticides, polychlorinated biphenyls, and metals at very high concentrations contaminate soil and groundwater. The EPA advised people against eating fish caught in the St. Johns River and nearby Deer Creek or drinking water taken from the local aquifer.

(HRS 70.71) Near the Village of DePue, Illinois, is a century old industrial site. The New Jersey Zinc Company began operations in 1903 on 175 acres of farmland. The facility grew over the years. The original plant produced slab zinc, used in the automobile and appliance industries, and sulfuric acid. They also produced zinc dust used in corrosive-resistant paints. During the height of operations, the smelter employed approximately 3,000 workers. In 1966, New Jersey Zinc constructed a diammonium phosphate (DAP) fertilizer plant. In 1972, Mobil Chemical Corporation took over operations. Since then, the EPA has identified elevated levels of metals, including zinc, lead, arsenic, cadmium, chromium, and copper spreading out from the site to residential properties within the Village of DePue. In addition, they documented contamination of a fishery, state wildlife refuge, and wetlands in and around Lake DePue.

(HRS 58.15) The small mining town of Picher, Oklahoma, sits amidst gigantic piles of mine waste covering some 25,000 acres. Established in 1913, the town became a mining center for lead and zinc in the 1920s and 1930s. When the mines closed in the early 1970s, ground water collected in the tunnels. In 1979, acid mine water containing high concentrations of lead, zinc, and other metals began discharging to the surface contaminating surface water. Acid mine water contaminated nearby Tar Creek so badly that the water turned red. A PBS documentary featured Picher, *The Creek Runs Red*. In 1981, the EPA designated 40 square miles around Picher and along Tar Creek a Superfund site. Oklahoma declared the area as its number one pollution problem. In 2006, the Army Corps of Engineers found most of the buildings in Picher in serious danger of caving in. A federal buyout program allowed most of them to move elsewhere, but some chose to stay behind despite the fact that there's no water and no police. In 2008, an F4 tornado hit the town destroying 150 buildings but it wasn't until September 1, 2009 that the town of Picher officially became a ghost town.

(HRS 54.66) Until banned in 1977, polychlorinated biphenyl (PCB) was widely used as dielectric and coolant fluids in transformers, capacitors, and electric motors. From 1947 to 1977, General Electric plants at Hudson Falls and Fort Edward discharged 1,100,000 lbs (500 metric tons) of PCBs into the Hudson River. For thirty years, the toxic chemicals accumulated in sediments that settled to the river bottom and in the bodies of all living things causing New York State to ban all fishing in the river. The contamination triggered a mutation in the Atlantic tomcod allowing them to survive but it didn't, however, prevent the tomcods from accumulating PCBs in their bodies and passing them on to striped bass and whatever else that eats them. Passing contamination from prey to predator is known as biomagnification. Even after thirty years, the PCBs remain a danger. In early August 2009, dredging along the upper Hudson River paused to let the sediment re-settle.[1] Seems the PCB levels rose above federal drinking water standards while they were playing in the muck.

Seen enough Superfund sites? I encourage you to go online and visit the EPA's National Priorities List for yourself. Find the closest Superfund site to you personally. Every state except North Dakota has them. Most sites fall into one of three categories, industrial, landfill or military base. My own Chandler, Arizona, has Williams Air Force Base (HRS 37.93) contaminated with a nice concoction of chromium, cadmium, lead, methyl ethyl ketone, toluene, polyurethane, paint thinners, jet fuel, lubricants, hydraulic fluid, and anything else you can think of from flight line and maintenance operations. Nice!

The National Priorities List tells only part of the story. In 2009, the EPA called for overhauling the laws governing toxic chemicals, specifically, the Toxic Substances Control Act of 1976. They proposed to include the 80,000 plus industrial compounds introduced since the statute took effect.[2]. Leaders of the chemical industry initially

1 8-12-09:Dredging of Hudson River resumes after forced stop, Wire Service, Arizona Republic
2 9-30-09: Obama aiming to reform statue on toxic chemicals, Wire Service, Arizona Republic

opposed any legislation but then reversed themselves and agreed that stricter federal regulation of their products was needed.[3] They admitted it is the best way to reassure consumers while they continue to make a pile of money. For over a year, the major players in the chemical industry adopted a public stance espousing reform just so they could have a seat at the table. What they really wanted was the opportunity to kill the bill. Millions of dollars later and out of the public's eye, their powerful lobbyists succeeded.[4] There were no new regulations.

In March 2010, the EPA announced that they want to impose stricter drinking water standards. At the heart of the proposal are stringent limits on four chemical compounds that we know cause cancer: tetrachloroethylene, trichloroethylene, acrylamide and epichlorohydrin.[5] A few days later, the EPA added perchlorate to the list, a rather nasty compound used in solid rocket propellant. Even small amounts can screw with your thyroid and cause a whole host of problems for developing fetuses.[6] These industrial chemicals are turning up in both ground and surface water all across America.

In September 2009, EPA announced new limits on three pesticides commonly used on western crops: chlorpyrifos, diazinon and malathion. Seems it interferes with salmon's sense of smell.[7] Don't worry. Unless you live near salmon waters in Washington, California, Oregon, and Idaho, they still use them on your food.

Three quarters of all American cornfields use a herbicide known as atrazine. According to Syngenta, the manufacture of the chemical for over 50 years, farmers around the world have relied on atrazine to fight weeds in corn, grain sorghum, sugar cane and other crops. Studies have shown that atrazine affects the natural hormone system in fish, birds, rats and frogs.[8] It has even caused male frogs to change gender. In March 2010, testing revealed atrazine in every major American and European river.

There are many modern chemicals turning up in our drinking water. Pharmaceuticals are the newest class of pollutants. Antibiotics, anti-depressants, birth control pills, seizure medication, cancer treatments, pain killers, tranquilizers and cholesterol-lowering compounds have all been detected in our water.[9] Nitro musks, used as a fragrance in many cosmetics, detergents, toiletries and other personal care products, are long-lived and bad for the environment. Some countries have banned nitro musks altogether. The active chemicals in your sun screen are building up in lakes and fish. The list is endless and growing.

Air pollution is also on our radar. In November 2009, scientists tracked huge concentrations of deadly atmospheric pollution driven eastward by the Earth's rotation.[10] American air pollution moves east across the Atlantic to Europe while Chinese pollution crosses the Pacific to America. Plumes of soot, ozone, mercury and a host of organic pollutants including DDT routinely travel from one continent to another.

For weeks in January 2010, schools all

3 8-9-09: Chemical industry backs tougher U.S. regulations, Wire Service, Arizona Republic
4 10-13-2010: Reform of Toxic Chemicals Law Collapses as Industry Flexes Its Muscles, Sheila Kaplan, Politics Daily
5 3-23-10: EPA to tighten standards on drinking-water purity, Wire Service, Arizona Republic
6 3-25-2010: EPA expected to set limits on additive in tap water, Wire Service, Arizona Republic
7 9-13-2009: EPA limits 3 pesticides that could harm salmon, Wire Service, Arizona Republic

8 3-2-2010: Herbicide found in water alters male frog hormones, Wire Service, Arizona Republic
9 Pharmaceuticals lurking in U.S. drinking water, MSNBC
10 11-10-09: Data track air pollution's transcontinental course, Wire Service, Arizona Republic

across Utah kept their students inside when air pollution skyrocketed to many times the federal standard.[11] Tiny flecks of pollution called PM2.5 swept through five counties. The EPA advised healthy people to reduce time spent outdoors and minimize their exertion.

In February 2010, Texas Republicans challenged the authority of the federal government and declared that greenhouse gases are not dangerous to people or the environment. They claimed the EPA based their findings on flawed science.[12] Ironically, Texas also leads the nation in greenhouse gas emissions and emphysema.

In March 2011, House Republicans proposed to strip the authority of the EPA to regulate greenhouse gas emissions from power plants, factories and mines. Thankfully, the bill died in the Senate.

A report came out in March 2010 stating that more Californians breathe air that fails to meet EPA standards than any other state in the union.[13] The non-profit RAND Corporation studied the cost of bad air in California from 2005 to 2007. They determined the cost to taxpayers at $193 million.

Dangerous chemicals and organic compounds get into our environment in a variety of ways and agriculture is one of the main sources. All across America, giant farms containing hundreds of thousands of pigs, chickens, or cows, produce vast amounts of manure. Sounds funny until you realize livestock waste seriously threatens our ecosystems. The problem has gotten worse as small family farms disappear into huge corporate farms run with brutal efficiency.

During the past 30 years, the number of hog farms in the United States dropped from 650,000 to 71,000, yet the number of hogs remained about the same.[14] In 1999, the U.S. Department of Agriculture reported that 2% of the hog farms in the country produce over 46% of the total number of hogs. Smithfield Foods operates eight slaughterhouses with combined capacity to process 110,000 hogs a day filling about 26% of American needs.

Brazil's JBS is the world's biggest beef processor, with capacity to slaughter about 51,000 head a day. As far back as 1996, 79% of cattle slaughters, approximately 22.6 million, occurred at only 22 plants.[15] In 2005, four companies controlled over 80% of America's beef and three of these same four companies along with an additional fourth processed over 60% of the country's pork.[16] Additionally, the same four companies processed over half of America's broiler chicken supply. It is the same situation for turkey meat. In fact, ten large companies produced more than 90 percent of America's poultry.

Smithfield Beef processes approximately 1.5 billion pounds of fresh beef annually with a processing capacity of 7,600 cattle per day. JBS Five Rivers is the largest cattle feedlot operation in the U.S. with an annual capacity of over 800,000 head of cattle in ten feedlots located in Colorado, Idaho, Kansas, Oklahoma, and Texas. The largest dairy farm in America is in Oregon and contains

11 1-13-10: Schools keep kids inside amid dirty-air warnings, Wire Service, Arizona Republic
12 2-17-10: Texas is challenging EPA on greenhouse-gas finding, Wire Service, Arizona Republic
13 3-3-10: Pollution-caused medical care cost California $193 million, Wire Service, Arizona Republic
14 1-13-2011: Agricultural Statistics Annual Report 2010, USDA
15 4-8-1999: Beefpacker Concentration, Mathews, Kenneth H. Jr., William F. Hahn, Kenneth E. Nelson, Lawrence A. Duewer, and Ronald A. Gustafson. USDA Economic Research Service
16 4- 13-2007: Concentration of Agricultural Markets, Mary and William Heffernan Hendrickson, Department of Rural Sociology, University of Missouri

40,000 dairy cattle. This consolidation has made the meat packing companies very powerful, while the government bodies that regulate them have done little to keep them in line. A typical mega farm will produce waste comparable to that of a small city with very little supervision.

Why should we care? These companies are entitled to make a profit. Locally they provide jobs and pay taxes, not to mention that they feed three hundred million Americans. Nevertheless, that doesn't mean we should turn a blind eye to how they go about their business. High levels of nitrates in drinking water increase the risk of methemoglobinemia, or "*blue-baby syndrome*," which kills infants. Animal waste also contains concentrated pathogens, such as *Salmonella, E. coli, Cryptosporidium*, and a host of fecal coliform. Manure transfers more than 40 diseases to humans, many of them fatal.

Manure from Wisconsin's dairy farms initiated the disastrous *Cryptosporidium* contamination of Milwaukee's drinking water in 1993, which killed more than 100 people, made 400,000 sick and resulted in $37 million in lost wages and productivity.[17] In 1996, the Centers for Disease Control established a link between spontaneous abortions and high nitrate levels in Indiana drinking water wells located close to feedlots.[18]

In Oklahoma, nitrates from Seaboard Farms' hog operations contaminated groundwater for many square miles, prompting the U.S. Environmental Protection Agency to fine the company $88,000 and issue an order requiring the company to provide safe drinking water to area residents.[19] In May 2000, 1,300 cases of gastroenteritis were reported and six people died

as the result of *E. coli* contaminating drinking water in Walkerton, Ontario. Health authorities determined the source was cattle manure runoff and cost taxpayers $155 million before it was over.[20] Iowa citizens took to the streets in 2010 to protest the easing of restrictions on the application of liquid hog manure on frozen ground. The measure passed anyway.[21] California officials have repeatedly identified manure as the major source of nitrate pollution in more than 100,000 square miles of polluted groundwater.[22] Yet, very little has been done to change the situation.

Another aspect of these mega farms is the forced evolution of viruses and bacteria. In 2009 alone, the huge conglomerates added over 29 million pounds of antibiotics to animal feed to

---

17 5-5-1993: 'Crypto' and controversy in Milwaukee water debacle, Ron Geiman, Wisconsin Light
18 7-5-1996: Spontaneous Abortions Possibly Related to Ingestion of Nitrate-Contaminated Well Water -- LaGrange County, Indiana, 1991-1994, Center for Disease Control

19 6-24-1998: Seaboard rides the wave, The Journal Record, Oklahoma City
20 12-20-2004: Canada's worst-ever E. coli contamination, CDC News
21 2-23-2010: Rally protests the easing of rules on hog manure, Wire Service, Arizona Republic
22 5-13-2010: Nitrate contamination spreading in California communities, Julia Scott, Press Enterprise

speed livestock growth. That's about 80% of America's total antibiotics use. This widespread use of antibiotics on animals contributes greatly to the rise of antibiotic resistant bacteria such as methicillin-resistant staphylococcus aureus (MRSA) making it harder to treat human illnesses.[23] Found in virtually every hospital and emergency room in America, now MRSA's in our locker rooms, gyms, and even public beaches.[24]

Hog manure emits hydrogen sulfide, a gas that causes flu-like symptoms in humans, but at high concentrations can lead to brain damage. In 1998, the National Institute of Health reported that 19 people died after inhaling hydrogen sulfide emissions from massive hog manure pits.

Not only do they smell bad, these huge open-air ponds filled with liquidized waste are prone to leaks and spills. In 1995, an eight-acre hog-waste lagoon in North Carolina burst, spilling 25 million gallons of manure into the New River. The spill killed about 10 million fish and closed 364,000 acres of coastal wetlands to shell fishing.[25] From 1995 to 1998, 1,000 spills or pollution incidents occurred at livestock feedlots across ten states and 200 manure spills resulted in the death of millions of fish and wide spread contamination. When Hurricane Floyd hit North Carolina in 1999, at least five manure lagoons burst and approximately 47 lagoons were completely flooded. Runoff of chicken and hog waste from factory farms in Maryland and North Carolina contributed to outbreaks of *Pfiesteria piscicida*, killing millions of fish and causing skin irritation, short-term memory loss and other cognitive problems in local people.

Storm runoff caries nutrients in animal waste out to sea where it causes algae blooms that deplete oxygen and create dead zones along our coastlines. Dead Zones are patches of ocean where there's not enough oxygen suspended in the water to support aquatic life. The size of the dead zone in the Gulf of Mexico fluctuates each year, extending to a record 8,500 square miles during the summer of 2002 and stretching over 7,700 square miles during the summer of 2010.[26] Since 2002, dead zones have occurred off the Oregon and Washington coast each summer to varying degrees causing massive fish die-offs.[27] In March 2011, millions of fish suffocated depositing a silver sheen of carcasses across Redondo Beach's King Harbor.[28] Plumes of ammonia, a toxic form of nitrogen gas formed during animal waste disposal, can travel hundreds of miles through the air before settling back onto the ground or the water where it starts another algae bloom.

This summary only scratches the surface of what we are doing to our planet. The average citizen does not have the time to research more than a few incidents, if any. We are too busy working for a living to do anything beyond what puts food on the table and pays the mortgage.

26 8-11-2010: Gulf of Mexico `Dead Zone' Grows as Spill Impact Is Studied, Leslie Patton, Bloomberg News
27 2008: Hypoxia off the Pacific Northwest Coast, Oregon State University, Partnership for Interdisciplinary Studies of Coastal Oceans
28 3-9-11: Oxygen dearth to blame for fish die-off in harbor, Wire Service, Arizona Republic

23 2-9-2010: Animal Antibiotic Overuse Hurting Humans?, Katie Couric, CBS News
24 9-13-2009: Resilient staph bacteria found on Washington (state) beaches, Wire Service, Arizona Republic
25 6-23-1995: Hog-waste spill fouls land, river in Onslow, Joby Warrick, *The News & Observer* of Raleigh, North Carolina

The Next Great Extinction

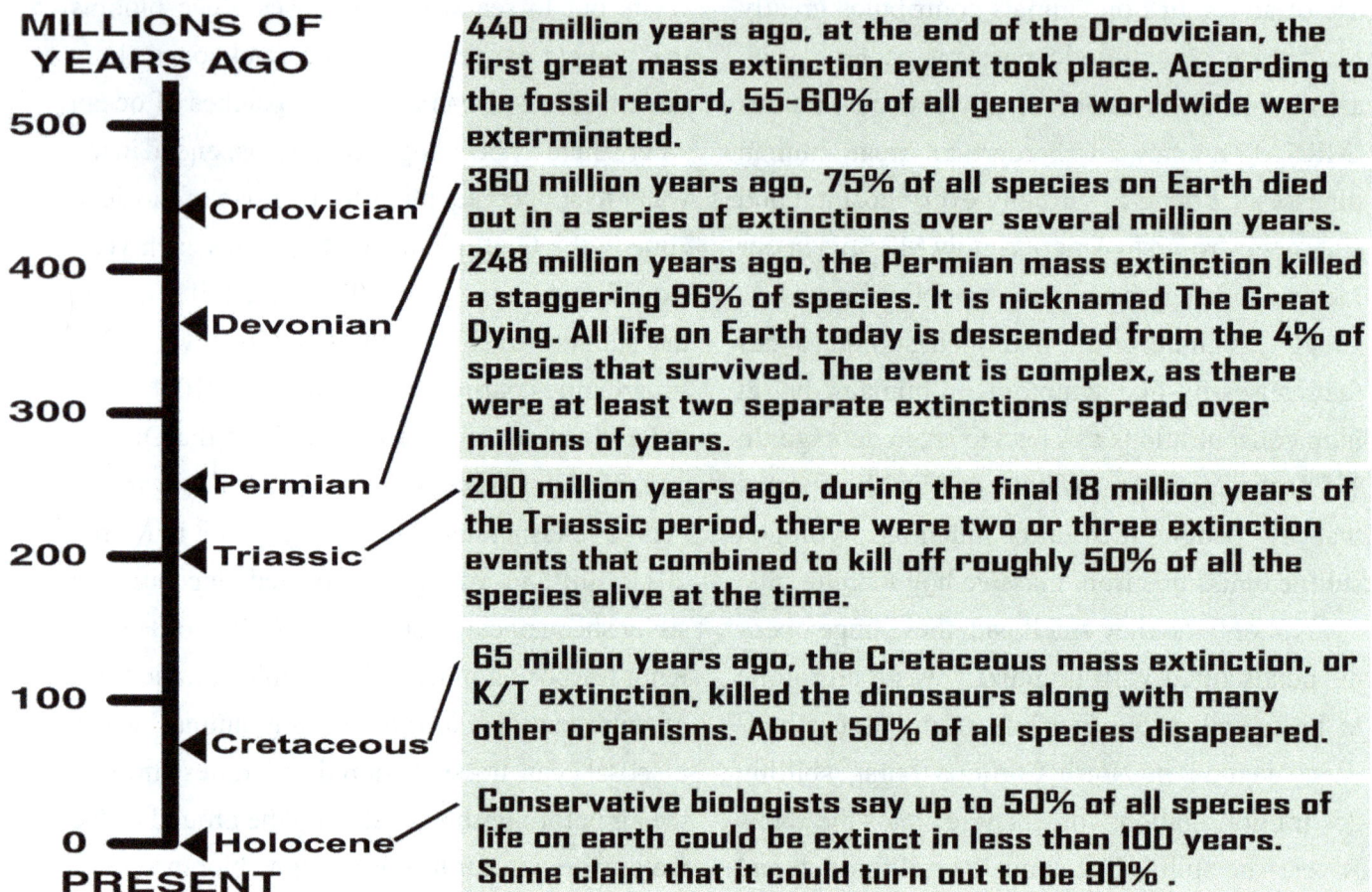

MILLIONS OF YEARS AGO

500 —

Ordovician

400 —

Devonian

300 —

Permian

200 — Triassic

100 —

Cretaceous

0 — Holocene

PRESENT

440 million years ago, at the end of the Ordovician, the first great mass extinction event took place. According to the fossil record, 55-60% of all genera worldwide were exterminated.

360 million years ago, 75% of all species on Earth died out in a series of extinctions over several million years.

248 million years ago, the Permian mass extinction killed a staggering 96% of species. It is nicknamed The Great Dying. All life on Earth today is descended from the 4% of species that survived. The event is complex, as there were at least two separate extinctions spread over millions of years.

200 million years ago, during the final 18 million years of the Triassic period, there were two or three extinction events that combined to kill off roughly 50% of all the species alive at the time.

65 million years ago, the Cretaceous mass extinction, or K/T extinction, killed the dinosaurs along with many other organisms. About 50% of all species disapeared.

Conservative biologists say up to 50% of all species of life on earth could be extinct in less than 100 years. Some claim that it could turn out to be 90% .

Climate scientists and energy speculators are not the only ones seeing a dark cloud on the horizon. Biologists are looking at the data and asking some very serious questions. We seem to be in the midst of a species die-off on a scale comparable with the mass extinctions of the geological past.[1]

Piecing together current events from all over the world and comparing it with the past paints a disturbing picture. This time, unlike the past, it is not a chance asteroid collision or a methane-spewing super-volcano that's at fault. Instead, it is us. The ever-growing human population is simply taking over the planet.

There are currently about 8.7 million species sharing planet Earth with us.[2] New species emerge and old ones fade away all the time, but normally on a time scale hard for humans to follow.

If extinctions happen all the time, as does the emergence of new species, why should we care? Biologists call the balance between extinction and emergence, the background level of extinction. If we count all species such as insects, bacteria, and fungi, not just the large vertebrates we are most familiar with, the background extinction is about one species per million species per year, or about nine species per year. In contrast, estimates based on deforestation rate indicate we are currently losing about 27,000 species per year to extinction from jungle habitats alone.

The typical rate of extinction differs for

1 3/3/2011:Study – Earth risks mass extinction, Wire Service, Arizona Republic

2 8/23/2011: How many species? A study says 8.7 million but it's tricky

different groups of animals. Biologists tell us the horseshoe crab has been around for a whopping 445 million years compared with mammals who exist for only about a million years, although some persist longer, some shorter. There are about 5,000 known mammalian species alive at present. Given the average species lifespan for mammals, the background extinction rate for this group would be approximately one species lost every 200 years. Of course, this is an average. The actual pattern of mammalian extinctions is much more uneven. Some centuries might see a dozen extinctions, and conversely, centuries will pass without the loss of any species. Yet, the past 400 years have seen 89 mammal species go extinct, almost 45 times the predicted rate, and another 169 become critically endangered.

Therein lies the concern biologists have for many of today's species. While the number of actual documented extinctions may not seem that high, scientists tell us that many more populations are so critically small that they have little hope of survival.

Other seemingly healthy species are in danger because of their symbiosis with another species. For example, the loss of honeybees can doom the plants they pollinate, and a prey species can take its predator with it into extinction. By some estimates, as much as 30 percent of the world's animals and plants could be extinct within 100 years. That's fast. These losses are likely to be unevenly distributed, as some geographic areas and some groups of organisms are more vulnerable to extinction than others. Species in tropical rainforests are at especially high risk, as are top carnivores, species with small geographic

ranges, and marine reef species. The reasons for this are predominantly human caused. We are stressing the world's biodiversity in many ways.

Destroying habitat and introducing creatures into ecosystems totally unequipped to handle them are the two biggest.

There's a major battle raging right now in the Midwest and our side is losing. The war started innocently enough. In the 1970s, fish farms in Louisiana and Mississippi imported Asian carp to clean aquaculture facilities and for fast-growing fresh meat.

That's the problem. They quickly turn into 4-foot, 100-pound plankton-eating monsters that devour 40 percent of their body weight every day thus stripping the environment of a key source of food. I wonder what genius thought this was a good idea. The fish soon escaped into the Mississippi River and have been spreading north ever since. They have breached every barrier we have put in their path and reached Lake Michigan the summer of 2010.[3] Fishermen in Lake Clumet, Illinois, just six miles from Lake Michigan, netted a 20-pound bighead carp. Where there is one, there are many more. A year later, academics, environmentalists and the fish industry called on the Army for help, specifically, the Army Corps of Engineers.[4] Unless someone takes radical action quickly, Asian carp will

---

3 6/24/2010: Asian Carp All Up in the Great Lakes
4 7/1/2011: Call in the army to protect Great Lakes from carp invasion says study,

decimate the Great Lakes native fish populations and the fishing industry. There's a new sheriff in town and his name is **CARP**.

Carp is not the only problem facing the Great Lakes. According to a local ecologist, the water in Lake Michigan is cleaner than it has ever been. The reason? An invasion of Quagga mussels, about 950 **trillion** of them, is sucking the life out of America's third largest lake. Quagga mussels are European shellfish with voracious appetites. Scientists estimate these mussels have reduced the biological activity in the lake by 30 percent. But that's not all. They have invaded the Southwest. In spite of our best efforts, the Quagga invasion reached Lake Mead and the lower Colorado River in early 2007.[5]

Maryland has outlawed felt wadding boots trying to contain the spread of rock snot in their trout streams. Rock snot is the thick mats of green algae that cover the stream bottoms.[6] The effort has only slowed down the invasion.

Pet pythons released into the wild more than a decade ago are growing to enormous size in the Florida Everglades eating everything including deer and alligators. These animals are now the top predator in their new environment. People have not fallen prey to these snakes yet but cats and dogs disappear from Florida yards all the time. I would worry about my kids if I lived near the park, especially the little ones. Park rangers estimate over 30,000 snakes infest the park and are struggling to contain them there.[7]

Species that compete with humans for food are all in decline. Bounties on bears[8], wolves[9] and mountain lions have all but eliminated them. The Eastern Cougar is extinct.[10] The Wolverine is extinct.[11] We kill sea lions because they eat salmon.[12] We protect our livestock at the expense of biodiversity.

Predators are especially at risk but species we love to eat are as well. Stocks of Atlantic Bluefin tuna, oceanic sharks and cod have been in decline for decades. Governments around the world are trying to slow down the slaughter and limit the number of fish taken each year but these species are highly migratory and cross

---

5 4/18/2011: Quagga mussels taking life out of Lake Michigan, Wire Service, Arizona Republic

6 3/22/2011: Stream anglers can't use felt-soled wading boots, Wire Service, Arizona Republic

7 10/31/2011: 16-Foot Python killed in Florida; Deer found in Stomach

8 6/17/2010: Alaskan official – Polar bear habitat hurts oil industry, Wire Service, Arizona Republic

9 8/25/2010: Suit by New Mexico ranchers targets gray wolf reentry, Wire Service, Arizona Republic

10 3/3/2011: US researchers declare eastern cougar extinct, Wire Service, Arizona Republic

11 3/16/2010: Only known wolverine in region is found dead, Wire Service, Arizona Republic

12 3/9/2010: First sea lion of the year is killed for eating salmon, Wire Service, Arizona Republic

Tsukiji Market-Tokyo

reproduce rapidly. They will inevitably crowd out specialized local species. Pigeons, zebra mussels, carp, rats, kudzu plants and tamarisk trees are examples of what biologists call "*weedy*" species, both animals and plants. Many weedy species will probably survive, and even thrive, in the face of the current mass extinction. However, hundreds of thousands, perhaps millions of others will perish.

What does this mean to us? Could we be on the path to oblivion? I think not. Humanity has proven itself a very persistent weedy species. We will survive all but the most catastrophic future. But do we want our descendants to pick through the rubble of a world that has lost much of its biodiversity? Once we lose that many species, the ecosystem will lose its ability to provide many of the valuable services that we take for granted, from cleaning the land, air and water, to pollinating crops, to providing a source for new pharmaceuticals. The fact is, we don't know what will happen when one species disappears, let alone so many. Which one will be the straw that breaks the camel's back? Which extinction event are we ignoring at our own peril?

many jurisdictions.[13] More than 40 fish species in the Mediterranean are in danger of immediate extinction due to overfishing, pollution and habitat loss.[14] A single 754 pound Bluefin tuna sold for a record $396,000 in January 2011 at an Asian fish market.[15] That's $525 per pound! How long can a species survive with that kind of price on their head? Biologists predict the worst. This just scratches the surface. Exploiting the creatures around us is our legacy.

One of humanity's biggest impacts on the extinction rate is habitat loss, an effect magnified by the burgeoning human population.[16] Now standing at 7 billion, the population of the world could double in 60 years if growth continues at the current pace. By draining wetlands, plowing prairies, spreading chemicals, logging forests, city and road building, we are altering the landscape on an unprecedented scale. Those organisms that do well under the conditions we've created tend to cope with change, tolerate a broad range of habitats, disperse widely, and

While the fossil record tells us that biodiversity has recovered after the most horrendous extinctions, it also tells us that the recovery will be unbearably slow in human terms, on the order of ten million years. Can we really afford to take that chance? Isn't it better to save what we have and not sacrifice it for short-term gain? **Space Based Solar Power** can help us do that.

13 3/18/2010: Stocks of tune and shark in sharp decline
14 4/19/2011: Many Mediterranean fish set to vanish, study says, Wire Service, Arizona Republic
15 1/5/2011: A Fish Story – Record Auction, ABC World News
16 12/2011: Landscape modification and habitat fragmentation – a Synthesis

61

The High Road

Why We Need Space More Than Space Needs Us

Only space has unlimited room for the seven billion people already here and those to follow in the coming generations. We are currently overwhelming the Earth's ability to support all of us.

For the benefit of those who skipped the Low Road and came straight to the good stuff, allow me to recap what you missed.

Unless we suffer a global disaster on the scale of the dinosaur-killing meteor, or we commit global thermonuclear suicide, sometime in this century the world's population will approach 10 billion. To think that Mother Earth can provide room and board for that many people is delusional. The electrical energy needs alone are overwhelming. Will we deny third-world citizens modern civilization? How can we justify stopping another country from developing a carbon-dirty, sulfur-rich coal-fired generator? Nuclear power is much worse. Once a country goes nuclear, it has some nasty stuff at its disposal. The proliferation of radioactive materials is as dangerous as that of nuclear weapons. It is just a matter of scale.

Over the course of the 21st century, climate change and rising sea levels will disrupt human activities and increase energy demands. While America and the European Union continue to consume the lion's share of Earths dwindling supply of hydrocarbons, China and India transform into industrial mega giants with a voracious appetite for energy. As the demand for energy rises, sources capable of meeting those demands are shrinking. In the years to come, we can expect rolling blackouts and brownouts like those suffered in California, not to mention that the costs will skyrocket. So, what are we doing to prepare for the coming energy shortage?

A close look at the various energy options currently available to us reveals a long list of inadequacies. Wind turbines and ground solar panels are too intermittent for heavy industry to depend upon. Hydroelectric is susceptible to drought and climate change not to mention the disruption a dam imposes on the environment. Bio-fuels compete with food production at a time when we have even more mouths to feed. All renewable energy sources have a place but they cannot sustain the base load electrical demand we currently have, let alone provide for future energy growth.

Only hydrocarbons are currently capable of meeting society's energy needs at the current level. Their use increased to record levels in 2011 and 2012 is on track to surpass this dubious achievement. Even knowing the harm it is doing to our environment, we increased the amount of burning with no end in sight.

While coal and oil have been a great boon to humanity, burning them with such complete abandon is accelerating global climate change and inducing environmental damage. In addition, hydrocarbons are a finite resource incapable of sustaining this output indefinitely and future economic growth will speed up the depletion rate.

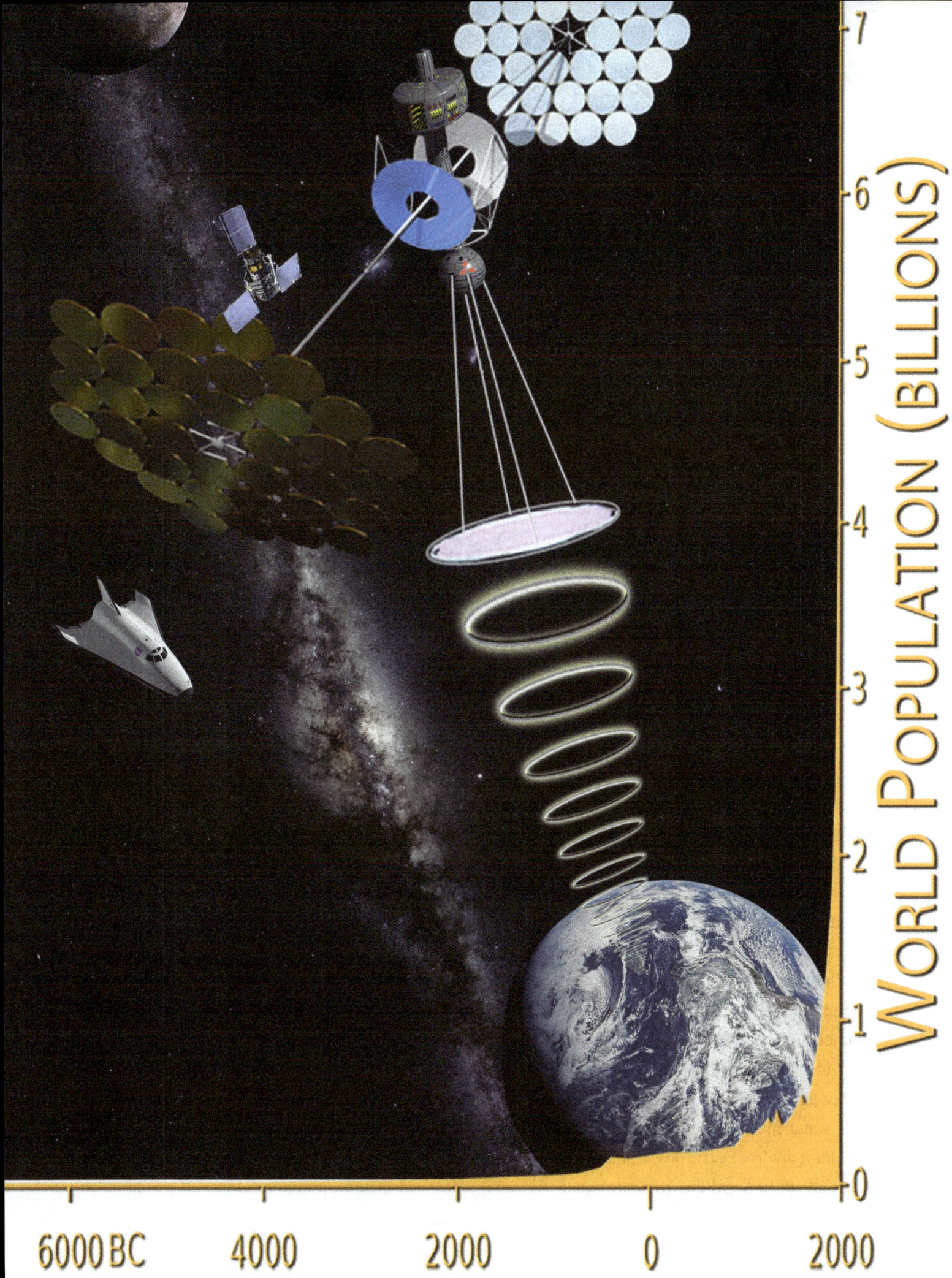

WORLD POPULATION (BILLIONS)

7

6

5

4

3

2

1

0

6000 BC   4000   2000   0   2000

Every new power plant that comes online brings us one-step closer to the end of easy energy. The world has already exceeded peak oil production and the only reason that natural gas hasn't, is the dangerous fracking process is forcing more methane out of the Earth even as it contaminates our drinking water supply. We still have coal in the ground for another 100 years but burning it spews contamination into our atmosphere that is wreaking havoc with our environment. We must find a better way to make electricity.

It has become abundantly clear that nuclear power plants are fraught with a host of problems. The number of plants the world would need to build is very high. Uranium mining itself contaminates the environment and purifying raw uranium is not exactly a clean process either. Spent fuel is highly toxic with no satisfactory long-term storage solution in sight. But the biggest nail in the nuclear coffin is the disastrous ramifications when something goes wrong and the unthinkable happens, radiation released into the environment, aka, Chernobyl, Three Mile Island, Fukushima, and I hope, no more. Even transporting radioactive material of any kind through our counties and states poses unacceptable risks. Did I mention nuclear plants cost a lot of money to build? Now, after Japan's experience with the Fukushima disaster, it is prohibitively expensive to even insure a nuclear reactor, let alone build one. I think the jury is in on nuclear energy, it is just too dangerous.

Water use is the sleeping giant in electrical energy generation. It is the largest consumer of fresh water in the United States and worldwide. Excluding hydroelectric power, the industry uses over 200 billion gallons of water each day to produce electricity. Of that, fresh water accounted for most of the total.[1]

How much energy does the world currently use? According to the U.S. Energy Information Administration,[2] the world used 505 Quad BTU in 2008. The same report projects an increase to 770 Quad BTU by 2035. They predict energy demand in China and India to double, while Western countries will increase 24%. Stated in purely electrical terms, the world in 2010 consumed 20.1 trillion kilowatt hours of energy. Of this, 13.9 trillion kilowatt-hours comes from burning hydrocarbons, 2.6 from nuclear, and 3.1 from hydroelectric. Earth-bound renewables hardly make a dent. The EIA predicts world net electrical production will increase to 25.5 trillion kilowatt hours by 2020 and 35.2 trillion kilowatt-hours by 2035. Even if you don't understand the enormity of the numbers, just remember that we are using more energy today than we did yesterday, by a lot. That trend will continue until something makes it stop.

There is only one power source capable of meeting this demand and all future demands, the *SUN*, and the best place to harness its power is collecting sunlight in *SPACE*, **Space Based Solar Power** or **SBSP**.

The simplest one-line definition of **Space Based Solar Power** is turning electricity generated from sunlight into microwave energy, beaming it through the atmosphere to receivers on the ground that convert it back into electricity that's fed into the existing power grid. Harvesting sunlight in space will become an economic pillar of the 21st century and far beyond.

In 1973, an engineer named Peter Glaser received patent number 3781647 on how to

---

1 Thermoelectric Power Water Use
2 International Energy Outlook 2011, EIA

convert solar radiation collected in space to electrical power down here on Earth.[3] That was almost 40 years ago. Our scientists have repeatedly studied the idea since then and believe we now have the technological proficiency to put it into place.

There are many advantages to **SBSP**. In space, there is no atmosphere to absorb the sunlight. Consequently, the intensity is much higher than any attainable on Earth's surface and remains virtually constant throughout the year. In space, there is no weather, no clouds, no night and day cycle, no earthquakes or tsunamis, even terrorists will have a difficult time getting at something in orbit.

Unlike coal, the energy **SBSP** provides is clean with negligible environment impact and can be delivered anywhere on Earth at a moment's notice. In fact, **SBSP** can respond quickly to demand, shifting resources to where they are needed most. Power shortages and brownouts will become a thing of the past.

The ground receiving antennas are simple and easy to build and can be designed for deployment on land or water. Floating power receivers allow the receiving antennas to be located offshore virtually anywhere in the world. Following the example of deep-water oil platforms, the power from dedicated receiving antennas can be used to electrolyze seawater on a scale that can supply the world with hydrogen. Think about it. **SBSP** has the potential to not only replace hydrocarbon and nuclear electrical power generation, but it can supply the energy to power a renewable hydrogen economy as well. There is no down side. This isn't a pipe dream. **SBSP** can and must be built. It is the future of the human race in more ways than just energy independence.

Desalination plants can supplement fresh water on a global scale and pumps powered by **SBSP** can deliver it where it is needed. Think of it, instead of oil pipelines running through your neighborhood, pipelines bring in fresh water from the oceans. A leaking pipeline would not be an environmental catastrophe. This is all possible if we only have the political will to colonize space.

◇◇◇◇◇◇◇◇◇◇◇◇◇◇◇◇◇◇◇◇◇◇◇◇◇◇◇◇◇◇◇◇◇◇◇◇

…Many years ago the great British explorer George Mallory, who was to die on Mount Everest, was asked why did he want to climb it. He said, "Because it is there." Well, space is there, and we're going to climb it, and the Moon and the planets are there, and new hopes for knowledge and peace are there. And, therefore, as we set sail we ask God's blessing on the most hazardous and dangerous and greatest adventure on which man has ever embarked.

President John F. Kennedy

◇◇◇◇◇◇◇◇◇◇◇◇◇◇◇◇◇◇◇◇◇◇◇◇◇◇◇◇◇◇◇◇◇◇◇◇

I admit that space colonization requires a tremendous investment in both people and treasure but look what we get for our money. A global effort will support a global economy for many centuries and America can lead the way. Space will inspire our young and much more.

Population pressure is stressing our planet to its breaking point but energy will ultimately decide if our civilization endures or was only a flash in the pan. Are we just an evolutionary aberration? Will humans disappear into the fog of time like so many other species? To keep this future from happening, we must expand our sphere of influence to include the Moon and the solar system now before it is too late. We must build our highway to the sky.

---

3 Method and apparatus for converting solar radiation to electrical power

Mother Earth and Father Sun

Ancient Egyptians had many gods and goddesses but the head honcho was Ra, their Sun God. Egyptians literally worshiped the Sun. Who can blame them? There it was blazing a path across the sky day after day without fail, it burnt your skin, raised your crops, and to look at it brought blindness. It is no wonder that a total solar eclipse terrified them and those who

Distance from Earth: 1.496 x 10⁸ km
(93,000,000 miles)

Intensity of Sunlight in Orbit: 1380-1400 W/m²
Sunlight at Earth's Surface: 0 - 900 W/m²

could predict such an event obviously knew the mind of God. These Bronze Age people believed that Nut, the Sky Goddess, swallowed Ra, the Sun God, at sunset and gave birth to Him every morning. Quite imaginative if you ask me. Truth is, they may have got the Sun part right.

Today we know Sun is not a god. Sun is a rather common star, one of billions like it in our galaxy, itself a mere mote of dust in the immensity of the universe. The truth behind Sun's existence is part of the grand story of modern science.

Physics has shown us that Sun is a million thermonuclear bombs going off every second for almost five billion years non-stop, burning with an intensity that warms our world even though it is 93 million miles away. The truth is, Sun is a gravity-constrained free-floating nuclear reactor and the ultimate power source at our disposal. Nothing else comes close. Because of Sun, our little world supports life, so in a way, Sun *is* god.

Either mankind will learn to collect the bounty that Sun sends us or we will wither and die on the vine. The way to do it only makes sense if we collect it in space. Our scientists and engineers have been studying the idea since 1968 when Dr. Peter Glaser first introduced the concept of using solar collectors in geosynchronous orbits to generate electricity from sunlight, use it to power a low-density microwave beam aimed at receiving antennas on the ground where it is then converted back into usable electricity and applied to the grid. Dr. Glaser's basic idea hasn't changed much over the years but our ability to accomplish it has. We now have the technology to put the first Powersat into orbit. Solar Power Satellite (SPS) or **Space Based Solar Power (SBSP)** can be built using current technology and a little engineering.

Throughout the 1970s and '80s, the Department of Energy and NASA examined the **SBSP** concept extensively, publishing several design and feasibility studies.[1] NASA updated

---

1 7/26/1971: Method and apparatus for converting solar radiation to electrical power – Peter Glaser, U.S. Patent 3,781,647

these studies three times in the 1990s and established the SSP Exploratory Research and Technology (SERT) program. NASA has been refining the idea for over 40 years identifying system concepts, architectures and technologies needed to produce a practical, economically viable source of electrical power that can take the place of coal, oil and nuclear. In 2007, the Pentagon's National Security Space Office (NSSO) got into the act, issuing a report stating that they intend to collect solar energy from space for use on Earth.[2] It was a matter of national security. They proposed spending $10 billion to construct a 10 MW Solar Power Satellite over a 10 year project. Their 10/10/10 plan would only demonstrate the science and develop the infrastructure for building much larger power satellites. Sounds like a bargain to me.

The National Space Society's Space Solar Power Library is an excellent accumulation of all this research.[3]

We are finally at a point in history when we *must* take **Space Based Solar Power** seriously. Burning hydrocarbons is polluting our world and nuclear power has shown its dark underside. **SBSP** is the only long-term means of producing electricity and space is the only reasonable place to collect it. Only **SBSP** still holds the promise of unlimited power without harming our planet. A thousand years from now Sun will still be shining. Will there be a drop of oil or chunk of coal left on Earth? Will abandoned uranium mines pockmark the land of our children's children, leaching uranium and radon into their environment? Or will our descendants enjoy a society that has fully embraced the life-giving energy of Father Sun. This generation is making that decision and what we decide will echo for centuries.

---

2 Space-Based Solar Power As an Opportunity for Strategic Security

3 Space Solar Power Library

Get Out of the Way

The island nation of Japan is the world's fourth largest economy but has very little domestic coal, oil or gas reserves. That doesn't stop Japan from burning through 185 million gallons every day making it the third largest oil consumer in the world. Even then, imported oil accounts for only 46% of Japan's energy needs.[1] Japan is also the world's largest importer of both coal (21%) and natural gas (17%).

The cost of importing fossil fuel into Japan has quadrupled in the last decade because like the U.S., Japan is addicted to energy. Lighted vending machines are everywhere. Toilets have heated seats with lids that flip up and down electronically. Gadgets abound and neon signs light up the night. Energy permeates Japanese society.

According to the U.S. Department of Energy, domestic sources account for only 15% of Japan's total energy needs and 11% of that comes from its wounded nuclear power program. The Ministry of Economy, Trade and Industry had planned to increase nuclear's share of total electrical generation from 34% in 2011 to 40% by 2017 and to 50% by 2030. Today, with the Fukushima disaster still unfolding, they are talking about how to replace their remaining nuclear reactors, the oldest first. If nuclear is no longer an option, what will take its place? More coal? More natural gas? There's only one thing that can supply over 225 GW of electrical power to the Japanese economy without harming the environment, **SBSP**.

The Japanese national space agency, JAXA, has been developing **Space Based Solar Power**

for more than a decade. Just in the past two years, Japan committed $21 billion towards building the first prototype. Before the Fukushima disaster, their goal was to bring the first 1 GW power satellite online by 2030. That date is no longer valid. They want to do it sooner.

JAXA researchers are working concurrently on two ground-based programs, one that uses microwaves and the other lasers, to transmit a kilowatt of power through the atmosphere. They will then do it from Low Earth Orbit (LEO). Ultimately, **SBSP** will introduce over 225 GW of clean solar energy into the country's power grid, so they want to get it right.

◇◇◇◇◇◇◇◇◇◇◇◇◇◇◇◇◇◇◇◇◇◇◇◇◇◇◇◇◇◇◇◇◇

"The technology for microwave transmission is more advanced, since it is based on current communication satellites."

Susumu Sasaki, Manager at JAXA's Advanced Mission Research Group

◇◇◇◇◇◇◇◇◇◇◇◇◇◇◇◇◇◇◇◇◇◇◇◇◇◇◇◇◇◇◇◇◇

Microwave beams can easily transmit huge amounts of power through the atmosphere but to keep the energy intensity down, the transmitting antenna in space needs to be big. The corresponding receiving rectenna on Earth needs to be even bigger.

Japanese scientists have been working on metal alloy plates that can directly convert sunlight into an infrared laser beam. The advantage is that the transmitting and receiving devices can be much smaller than for microwaves. Also, lasers will not interfere with microwave communication networks.

---

[1] 3/2011: Japan – Country Analysis Briefs, EIA/DOE

However, unlike microwaves, clouds block lasers and the atmosphere absorbs a large portion of laser energy. Besides, thanks to science fiction, most people think space lasers are weapons. JAXA scientists assure us that there are ways to guard against that but it is a tough sell against the virtually harmless microwave beams. A person can safely walk through the center of a microwave power beam and at most, feel it warm their skin reminiscent of a sunny day. Japan is confident they can make the system work.

100MW Ningxia Shizuishan Solar Power Plant, China-1

The gossamer nature of these designs make it feasible to launch the Megawatt prototypes from Earth but eventually, as the science and engineering become better defined, the Gigawatt power satellites that will replace coal must be constructed using materials we harvest from the Moon. Japan is taking the minimalist approach, lift from Earth only what you need to mine the metals and manufacture the solar cells from lunar ore. Don't attempt to muscle billions of tons up Earth's gravity well.

China is playing it close to the vest but their intentions are clear; they intend to conquer space and seize this cheap energy source. Recent successful docking missions pave the way for a Chinese space station and what plans we know of include colonizing the Moon and developing **SBSP**.[2] China can surpass the United States in space only if America stops competing.

In recent years, there has been a steady stream of announcements from private companies and international space agencies seeking to be the first to master **SBSP**. In 2008, long before the Fukushima Disaster, Japan announced its **SBSP** program goals. In 2009, the giant California utility company Pacific Gas & Electric contracted with Solaren Corporation to deliver 200 MW of power from space by 2016. In 2010, Astrium, the space-exploration subsidiary of the European Aeronautics and Space Exploration Company (EADS) announced their intentions to build a 20 KW prototype by 2015.

◇◇◇◇◇◇◇◇◇◇◇◇◇◇◇◇◇◇◇◇◇◇◇◇◇◇◇◇◇◇◇◇◇◇◇◇

"We choose to go to the Moon in this decade, not because it is easy, but because it is hard."
President John F. Kennedy

◇◇◇◇◇◇◇◇◇◇◇◇◇◇◇◇◇◇◇◇◇◇◇◇◇◇◇◇◇◇◇◇◇◇◇◇

The United States readily achieved that objective and, effectively, won the Cold War. A similar challenge now faces us in the race for **Space Based Solar Power**.

---

2 11/19/2011: China calls docking mission big step toward
   space station, AP

National Security

Controlling real estate has always been a major objective of a military force. Besides killing enemy combatants, it is perhaps the most important factor in winning a battle, and nothing is more critical in controlling an area than the ability to move your forces about it quickly. Chariots gave advantage to the Egyptians and mounted soldiers have turned the tide from Genghis Khan to the Battle of Gettysburg. The ability to move forces across the battlefield is paramount. World War I effectively ended the horse as a weapon of war. Machines have taken its place. In 1919, a transcontinental road trip planted a seed in the mind of a young Army officer crossing America as part of a military convoy. The route was the Lincoln Highway. The officer was Lieutenant Dwight D. Eisenhower.

"Though force can protect in emergency, only justice, fairness, consideration and cooperation can finally lead men to the dawn of eternal peace."

Dwight D. Eisenhower

Twenty-five years later, that Lieutenant became General Eisenhower, Supreme Commander of the Allied forces fighting the Nazis. While in Europe, he witnessed firsthand the advantages the Autobahn gave to the German national defense system. In 1956, President Eisenhower put that lesson to use when he championed, then signed into law, the National Interstate and Defense Highways Act.

He believed the project was essential to national security. He argued that if American cities were threatened, the interstate highways could quickly evacuate them and allow the military to move in. Construction took 35 years at an estimated cost of $425 billion in 2006 dollars making it the largest public works project in American history. As of 2006, it had a total length of 46,876 miles (75,439 km) and growing.

The Interstate Highway System is the backbone of America and one of the defining elements of our civilization. Millions of tons of food and manufactured goods move over this transportation network every day, not to mention us. Who hasn't driven on the interstate? America would be very different without them. Construction and maintenance of our roads certainly generates jobs but the real advantage the interstate system gives all of us is providing the infrastructure that sustains our way of life. Here in Arizona, I have witnessed the urban sprawl follow along the track of a newly constructed freeway. Homes and businesses, entire neighborhoods spring up where before was empty desert. The phenomenon is also apparent along the new Valley Metro Light Rail in the Phoenix metro area. The rail corridor passes through many cities and everywhere it goes, prosperity follows. Infrastructure projects go far beyond mere construction. Their presence revitalizes neighborhoods and provides a framework for modern civilization. It is in our national interests to extend this infrastructure into space.

America's military has already concluded that it is a matter of national security. In 2007,

the Pentagon's National Security Space Office released a report titled **Space-Based Solar Power as an Opportunity for Strategic Security**. The report states that **SBSP** can advance American security and merits significant additional study. Specifically, the report calls for the U.S. Congress to appropriate $10 billion over 10 years culminating with building a prototype in geosynchronous orbit capable of beaming 10 megawatts of power to a receiving station on the ground. 10/10/10 is a modest proposal, a fraction of the cost of our Interstate Highway system. We need a president like Dwight D. Eisenhower to champion this issue, to make America see the enormous benefit that space colonization will bring. Will Obama have the courage? The amount of money involved is enormous but the benefit to humanity is impossible to overstate.

Back in 2004, when President George W. Bush laid out his *Vision for Space Exploration*, he included a call to develop off-planet resources, specifically, the Moon. Our nearest neighbor contains the raw materials needed to build just about anything we want to build, not to mention that it is right under our noses. During the next full Moon, go outside and stare at it for a while. Take your time and think about this… It is entirely possible to do a bunch of work on the Moon with robotic machines teleoperated from Earth without sending people there at all. We already know how to do it. The Mar's rovers did pretty well and if we can teleoperate machines on Mars, then we sure can do it on the Moon. Don't take this wrong. People will eventually go there but first we must build the infrastructure to support living creatures using machines.

The plan NASA produced to fulfill President Bush's vision is the *Exploration Systems Architecture Study* (ESAS). After reviewing this 758-page document, it struck me how much it resembled the Apollo Program. Sure, they used redesigned Shuttle hardware but essentially, it was a series of Moon shots just like the old program. Project Constellation quickly became a straightjacket and in 2010, with the project vastly over budget, President Obama cancelled it.

With the completion of the International Space Station (ISS) triggering the retirement of the Space Shuttle, suddenly America doesn't have a manned space program anymore. We are dependent upon the Russians, Europeans, and Japanese to resupply the station. American cargo launchers have yet to deliver a single pound to the ISS let alone change the crew.

A different approach is one proposed by Dr. Paul Spudis of the Lunar and Planetary Institute in Houston, Texas and his colleague Tony Lavoie of NASA's Marshall Space Flight Center in Huntsville Alabama. In December 2010, they released a nice little report, only 30 pages, titled *Mission and Implementation of an Affordable Lunar Return*. In it, instead of rehashing the Apollo Program, they propose to establish the infrastructure to harvest water off the Moon using robots teleoperated from Earth. Their plan is affordable, flexible and not tied to any specific launcher. Individual pieces are small, permitting them to be deployed separately on small launchers or combined together on single large launchers. The goal is a fully functional, teleoperated lunar outpost capable of producing 150 metric tons of water per year. That's enough water to create a cislunar transportation system capable of supporting the building of **Space**

Based Solar Power, among many other things.

This plan is much more in line with what both President Bush and President Obama had in mind. Their objective is not a series of Apollo-style expeditions but rather something more ambitious and permanent. The goal is nothing less than the inclusion of the Moon into the human sphere of influence. The high cost of launch to orbit is a formidable barrier to overcome. However, despite numerous and continued attempts to lower launch costs over the last 30 years, it varies from $5000/kg to over $21,000/kg. Most discussions accept the average of $12,000/kg. Launch cost is a *Catch-22* problem: costs are high because volume is low and volume is low because costs are high. In the future, we may expect to see launch costs drop by factors of two or three but more is unlikely.

We have the rockets. The challenge is to engineer an architecture that accomplishes something that's never been done and do it with less money. Human beings require a lot of life support so by necessity, manned missions will be restricted to Low Earth Orbit (LEO). This puts a heavy reliance on robotics and remote operating for the work taking place on the Moon and cislunar space. Yet, if we attack the problem in small, discrete steps, we will eventually climb the mountain and build our highway to the sky.

People today argue against returning to the Moon. *Why should we go back? What was the justification for spending all that money on the Apollo Program to begin with?* All we got in return was a few lumps of Moon rock. Aren't there better causes here on Earth? These are all good questions that deserve good answers.

In a way, the situation is like that in Europe before 1492. Sailors knew how to reach Asia sailing around Africa or overland through the Middle East but some captains were convinced they could do the same thing by sailing due west across the Atlantic. Christopher Columbus was one. Columbus first approached King Henry of Portugal for ships to discover the way to China and Japan but the King said no. His people advised him not to waste money on such a boondoggle.

Then Columbus went to Spain and asked Queen Isabella. "*Wait until the war is over*," the Queen told him.

Columbus waited. When the war was finally over, in 1492, he came back and asked again. This time Queen Isabella said yes against the recommendations of her advisors.

Today, the naysayers argue that space is a waste of money. Yet the discovery of North America made a profound difference to the history of Western Civilization. Extending our reach to the Moon will have an even greater effect and is well worth whatever it will take. It will completely change the future of the human race and may determine whether we have any future at all. The colonization of space and the Moon won't solve all of our problems on planet Earth, but it will solve some of the biggest and give us a new perspective on all of them. It will force us to look outwards, rather than inwards.

Hopefully, it will unite us to face the common challenge. Colonization is a long-term strategy best served by joining many nations together. We could have human-occupied communities on the Moon within twenty years, reach Mars in thirty years and send explorers to the outer planets by the middle of this century. The best way to make such a journey is one step at a time.

Space Based Solar Power

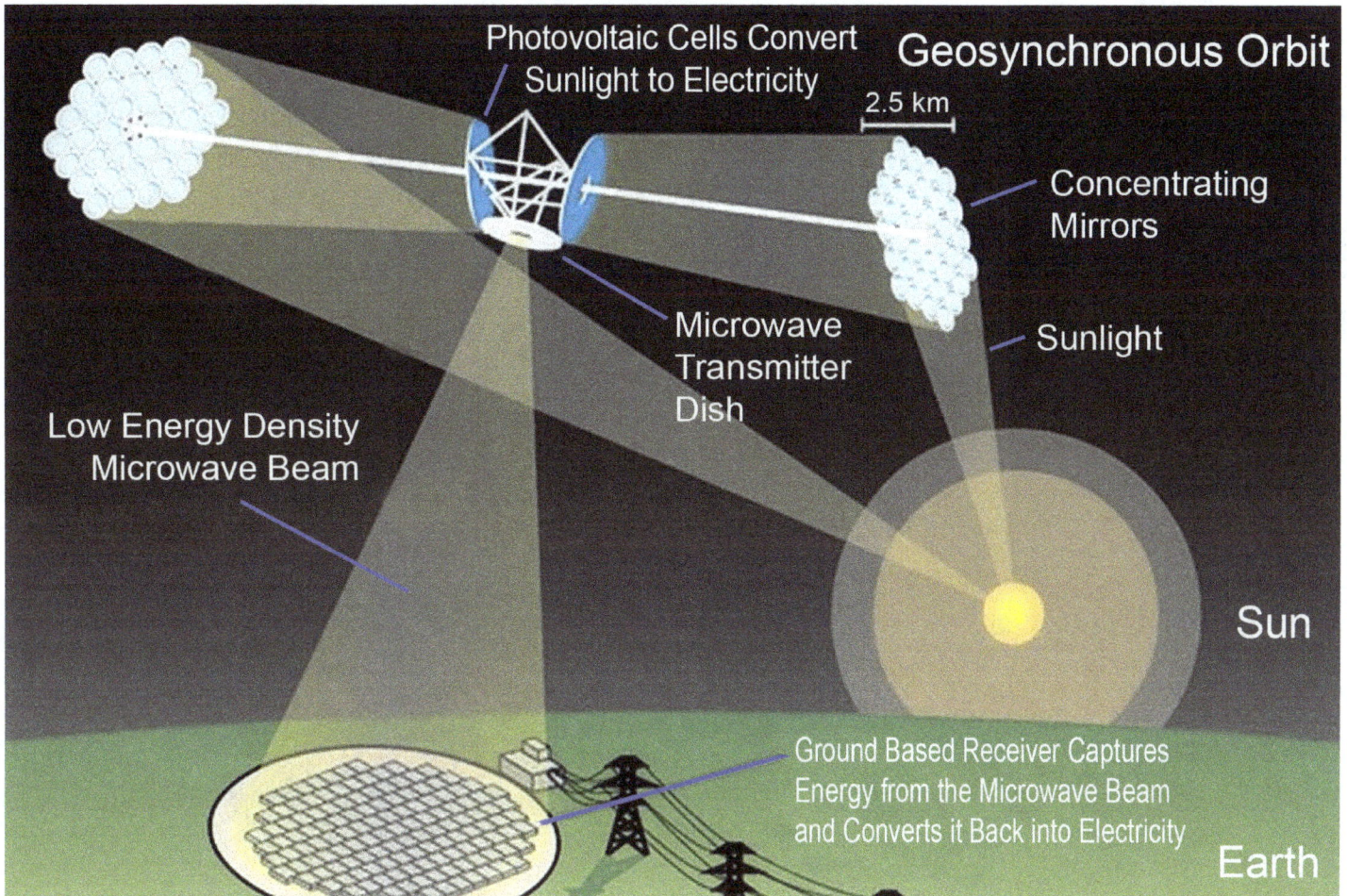

Image courtesy of New Scientist. Sunlight is reflected off giant orbiting mirrors to solar collectors where it's converted to electricity, changed into microwaves, and beamed to earth. Ground-based rectennas capture the microwaves and convert them back to electricity, which is supplied to the grid.

For those who have never heard of **Space Based Solar Power** (**SBSP**), let's layout the basics. **SBSP** is composed of three major elements, two of them in space and the third on Earth's surface. The collector and transmitter are separate satellites but many depictions show them tethered together. In this discussion, the actual design is not as important as the technology behind it. We can do this.

In approximate numbers, it takes a 3.3 GW power station to supply electricity to about 4 million people. If we calculate how many 3.3 GW

powersats would be needed to supply 7 billion people, it works out to be 1,750. Any thought of lifting that much mass off the surface of the Earth using rockets should be gone. Mining the Moon is the only answer and even that will take a while. My grand kids will watch **SBSP** grow and hopefully, actively participate in its construction. **SBSP** will become the backbone of a new space faring civilization, very similar to what the interstate highway system did for America.

Lets' first look briefly at a summary of the three parts then go into detail about them.

Solar Collector

The solar collector is in orbit where it receives raw sunlight and converts it into electricity. Some designs combine reflectors with solar cells to maximize the energy collected but it is essentially the same technology used on our existing fleet of satellites or even on the roof of your home only on a much larger scale.

Microwave Transmitter

Similar to the technology we use in our cell phones, the transmitter converts the electricity generated from the solar collectors into microwave energy and beams it down to the surface of the Earth. The beam's frequency determines its susceptibility to atmospheric absorption and loss of power. The beam's energy density is kept low by making the beam's cross section large.

Ground Rectenna

Using common electrical components, the rectenna is a simple array that directly converts the incoming microwaves back to electricity and feeds it into our existing power grid. The size of the array must be large to accommodate the incoming beam.

Harvesting Sunlight

That's it. Simple, right? But everyone knows the devil's in the details so let's begin at the top. The power density, also known as the solar constant, is a measurement of the energy coming from the Sun. Since the Earth is about 93 million miles from the Sun, by the time the sunlight gets here, the power density has dropped to approximately 1400 watts per square meter (W/m^2) in space, but due to atmospheric absorption, less than 1000

W/m^2 reaches the Earth's equatorial surface on a cloudless day at high noon. That's best case. It goes down from there. Ground solar power not only suffers from atmospheric absorption, but cloud reflection, time of day (zero at night), and latitude as well. In space, we can harvest energy 24 hours a day, 365 days a year without regard

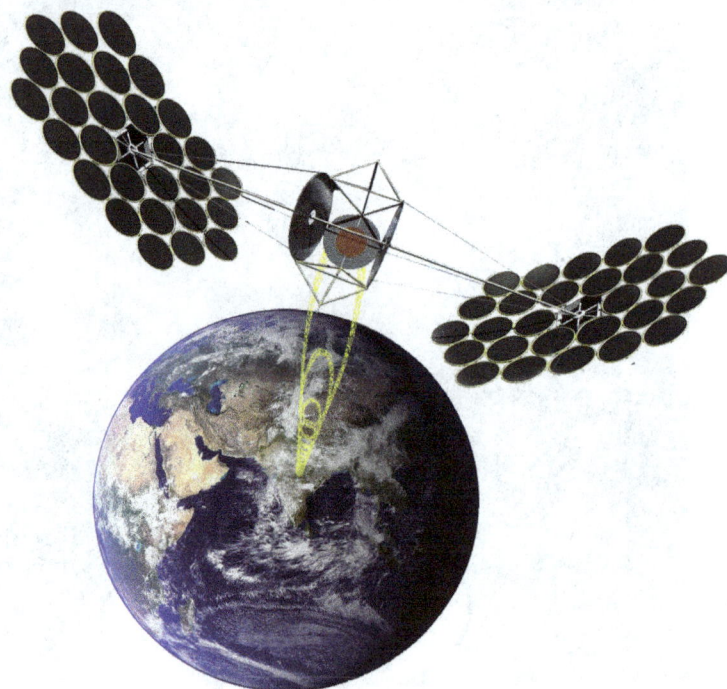

for storms or earthly disasters. However, to do so we need to construct great arrays of solar cells in space, or a combination of solar cells and reflectors, to convert the abundant sunlight into electricity.

There are many solar collectors already in space. On March 17, 1958, the U.S. Navy launched Vanguard 1. It was the fourth satellite ever launched and the first to be solar powered. Since then, the space industry has placed over 3000 satellites in orbit, virtually all of them powered by the Sun. In 2010, solar powered satellites were a $160 billion a year industry.[1]

1 6/8/2010: 2010 State of the Satellite Industry Report – SIA Satellite Industry Association

The biggest man-made satellite to date is the International Space Station with eight large, wing-like solar arrays, each measuring 112 feet long and 39 feet wide (34 m x 11 m). The arrays together contain 262,400 solar cells and cover an area of about 27,000 ft² (2,500 m²). A computer-controlled gimbal rotates to keep them tilted toward the Sun as the ISS zips around the Earth every 92 minutes.[2]

A typical satellite needs a few tens of kilowatts of energy at most. The ISS array generates 110 kW, but a single utility-grade Solar Power Satellite (powersat) will produce much more, enough to power a small city. Simply scaling up the ISS solar arrays will not suffice. Conventional solar cells are too heavy and inefficient. We need something lighter.

Amorphous silicon absorbs solar radiation 40 times more efficiently than single-crystal silicon and a film only 1-micrometer (one-millionth of a meter) thick can absorb 90 percent of the usable light energy shining on it. Vanguard Space Technologies is developing the material into high-efficiency thin-film solar cells with integral flex-circuit wiring that provides ultra-high specific power, more than 400 W/kg.[3]

In a separate project, NASA's Glenn Research Center contracted with PowerFilm Solar of Boone, Iowa, to develop amorphous silicon thin-film solar cells for large space applications. They came up with thin, flexible solar panels manufactured using a proprietary, low-cost, roll-to-roll, monolithic process that eliminates the need for manual connecting individual solar cells.[4]

Another breakthrough came in early 2010 when a team of researchers at the California Institute of Technology (Caltech) announced that they have created a new type of flexible solar cell also manufactured in rolls. They imbedded fine amorphous silicon wires in a polymer, and then brought them together in a continuous array. The researchers found the wires interact to increase the cell's ability to absorb light. So much so that the new solar cells have surpassed the conventional light-trapping limit for amorphous materials, absorbing up to 96% of incident sunlight at a single wavelength and 85% of total

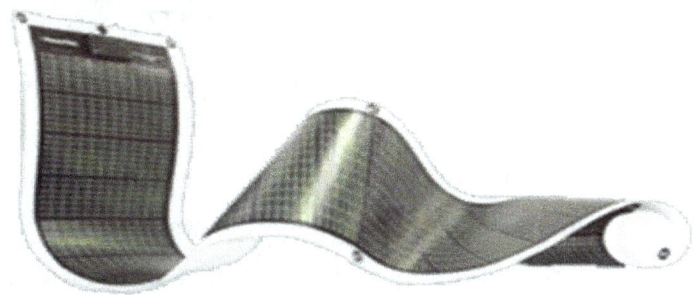

collectible sunlight, converting between 90% and 100% of the photons they absorb into electrons. In scientific terms, the wires have near-perfect internal quantum efficiency.[5]

Placing solar cells on flexible materials provides **SBSP** with a low weight, attractively priced solution to the tremendous acreage requirements of a Space Based Solar Collector.[6] For those without an engineering degree, being able to achieve a high power-to-weight ratio is the crown jewel of **Space Based Solar Power**. The only thing better would be if we could use lunar materials to make the rolls! Lifting them from the Moon is much easier than lifting them from Earth.

A circular solar collector one kilometer in diameter has over a gigawatt of solar energy fall upon it in orbit. Using conservative numbers, the new flexible solar rolls will deliver over 850

2 International Space Station – Facts and Figures
3 Space Power and Thermal Control
4 Paper-Thin Plastic Film Soaks Up Sun to Create Solar Energy, NASA Spinoff
5 2/17/2010: Less is more for highly absorbing, flexible, cheaper solar cells, Darren Quick
6 Roll Solar Panel

SBSP Microwave Power Density Characteristics at Rectenna Ground Sites

Sunlight = 140 mW/cm²

0.02 mW/cm²
0.08 mW/cm²
0.10 mW/cm²
1.00 mW/cm²

Peak Power Density = 23 mW/cm² at the center of the beam

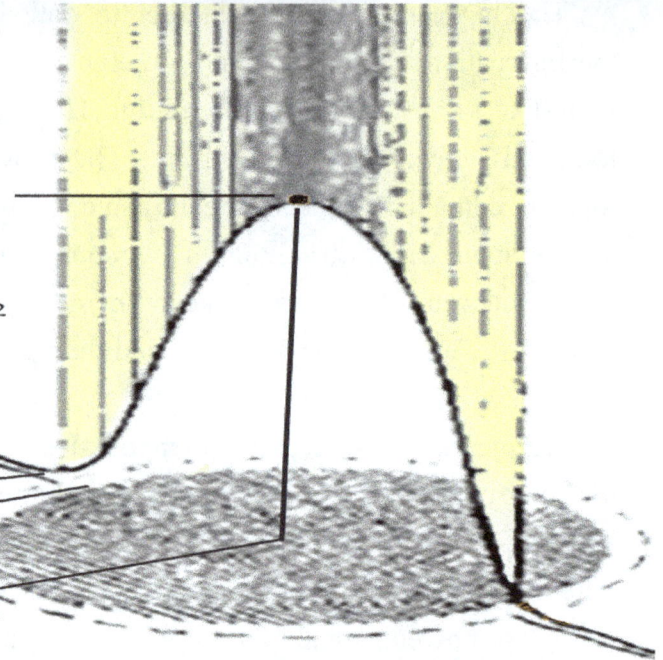

MW of electrical power to the microwave beam. Even after beaming through the atmosphere and yet another energy conversion, that will translate into over 600 MW delivered into our existing power grid 365/24/7. That is only the beginning. We will not stop at one.

Wireless Transmission

It takes a unidirectional microwave antenna to convert the electricity from the solar collectors to microwaves and beam it down to the Earth's surface. To keep the overall energy density low, physics dictate a transmitting antenna at least 1000 meters in diameter. The largest antenna on Earth is the Arecibo radio telescope in Puerto Rico at 305 meters and massing thousands of tons. However, one of the biggest advantages to building in orbital space is zero gravity. Where Arecibo has heavy metal panels and I-beams, its space-based counterpart is vast sheets of one millimeter thick metalized Mylar and tethers. A space antenna needs very little supporting structure to function, more like a kite.

Most people cannot get past the microwave beam component of **SBSP** so let's talk about it. To them, a microwave oven heats up their cup of coffee, so common sense tells them it must do the same to our atmosphere, cooking any bird unfortunate enough to fly through the beam. I assure you, that's not the case. Let's start with a little history.

Wireless power transmission started in 1975 when NASA JPL Goldstone transmitted 34,000 watts across a mile with 82% conversion efficiency.[7] Over the next five years, the U.S. Congress authorized NASA and the Department of Energy to develop the technology.[8] Funding stopped when the Reagan Administration assumed power in January 1981. In fact, most research into solar power came to a screeching halt. Oil and coal were cheap again.

It was almost thirty years later, in 2008, when the Discovery Channel sponsored an experiment

7 1975 NASA JPL Goldstone Demo of Wireless Power Transmission, YouTube
8 October 1978: Satellite Power System Concept Development and Evaluation Program Reference System Report. DOE/ER-0023,322

demonstrating power beaming across 92 miles of dense atmosphere.[9] They beamed power from a mountaintop in Maui to the main island of Hawaii. With the support of scientists in Japan, Texas and California, former NASA physicist John Mankins spent less than $1 million over five months reaffirming that the wireless transmission of power concept is sound. They used mirrors to focus sunlight on some new and improved solar cells able to get five times more electricity than previous cells. The amount of energy they beamed was modest, only 20 W, but even using portable equipment (they had to disassemble it every night because the mountain they were on is sacred), they achieved a system-wide calculated efficiency of 64%.

In nice round numbers, the average energy density of sunlight in geosynchronous orbit is 1400 W/m^2. That works out to be 140 milliwatts per square centimeter (mW/cm^2). The energy density in the proposed **SBSP** beam reaches a maximum of 23 mW/cm^2 in the geometric center of the beam. A one gigawatt power beam would require a receiving rectenna about 2.5 kilometers in diameter depending on its latitude. Outside the beam, the density quickly drops to zero. Yet, the microwave beam passing through the atmosphere is always a point of contention for skeptics of **SBSP**. They fear that the beam itself harms the environment or worse, that it could be turned into a weapon.

Coming back to your microwave oven, it just makes sense that if microwaves can heat up my cup of coffee, they must heat up the air as they pass through the atmosphere. Right?

9 9/12/2008: Researchers Beam Space Solar Power in Hawaii, Wired Science http://www.wired.com/wiredscience/2008/09/visionary-beams/

Almost right. The energy density inside your microwave is many orders of magnitude greater than the proposed power transfer beam and yet, it takes a full minute to warm your cup of coffee. Microwave energy is non-ionizing, that is, it does not carry enough energy to damage living tissue even at the levels inside your oven.

Since 1921, when Albert Wallace Hull invented the magnetron, scientific studies have sought to find any repeatable evidence of microwave's adverse effects on living tissue. Even high-intensity microwaves have just enough energy to vibrate atoms within a molecule, but not enough to remove electrons and create an ion. Examples of this kind of radiation are sound waves, visible light, and yes, the microwaves in your oven. Microwave radiation is absorbed near the skin's surface and the only way it can damage tissue is through heating. That requires high-density microwaves over a significant period.

Microwave radiation, like sunlight, is all

PowerSat 2.45 GHz

1 Hz			1 MHz										
1.0	10^2	10^4	10^6	10^8	10^{10}	10^{12}	10^{14}	10^{16}	10^{18}	10^{20}	10^{22}	10^2	

Frequency (exponential scale)

long radio waves   short radio waves   infrared   ultraviolet x-rays   gamma rays

visible
light

$$f = \frac{c}{\lambda} \qquad \lambda = \frac{c}{f}$$

f is frequency in cycles
   per second (Hz)
c is speed of light
   299,792,458 m/sec
λ is the Greek leter Lamda
   and represents
   wavelength in meters

AM
radio

FM
TV

microwaves

Wavelength (logarithmic scale)

10^8	10^6	10^4	10^2	1.0	10^{-2}	10^{-4}	10^{-6}	10^{-8}	10^{-10}	10^{-12}	10^{-14}	10
		1 km		1 m		1 cm	1 um	1 nm				

PowerSat 12.2 cm

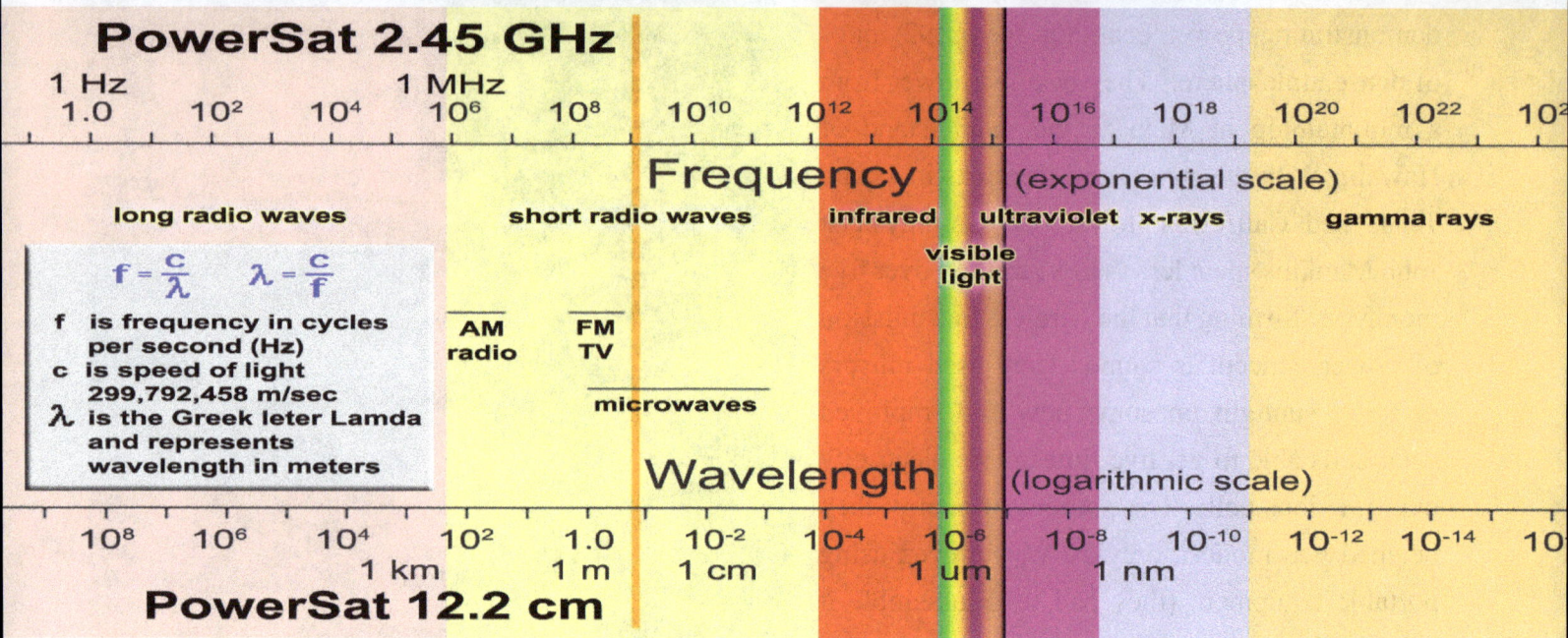

around us. Our personal cell phones, radio stations, airport radar, even some TV remote's use microwaves to communicate. Many industries such as plastics and paper use millions of industrial microwave heaters and dryers every day. Citizens band radios transmit microwaves at 4,000 milliwatts (4 watts) in all directions. All this, yet no reputable scientific study has ever turned up anything alarming.

The frequency of the proposed energy beam is 2.45 GHz. Interaction with the atmosphere is negligible at this wavelength. Clouds and weather become transparent to the beam.

Don't be misled by the sensationalist writings such as *The Zapping of America* by Steven Broder.[10] Since its publication over three decades ago, not a single peer-reviewed study has shown any harmful effects attributable to microwaves, and believe me, people have tried. These efforts have shown just the opposite, that microwaves are harmless. As far back as 1983, the EPA studied the effects of microwaves on rats and found no

significant effects at 2.45 GHz[11] at considerably higher densities than the proposed power beam. The Food and Drug Administration (FDA) says that the tremendous weight of scientific evidence has not linked cell phones with *ANY* health problems.[12] A National Cancer Institute study found that, despite the dramatic increase in cell phone use, occurrences of brain cancer did *NOT* increase between 1987 and 2005.[13] Study after study will fill a small library. Finally, on May 17, 2010, the World Health Organization (WHO) released the results of the largest study yet on the effect of microwaves on our brains, finding *NO* increased health risk due to microwave energy emitted by cell phones.[14]

These studies are not the only ones. Numerous other reports exist on birds roosting and nesting in high power microwave radar and communications antennas, including the horn, with no apparent ill effects at many times the

10 October 1977: W W Norton & Co Inc.: Zapping of America: Microwaves, Their Deadly Risk, and the Coverup - by Steven Broder

11 1/1983: Effects of 2450 MHz on Cerebral Energy Metabolism - EPA
12 5/18/2010: Do cell phones pose a health hazard? FDA
13 5/17/2010: International Study Shows No Increased Risk of Brain Tumors from Cell Phone Use, National Cancer Institute
14 No evidence linking cell phone use risk brain tumors, WebMD

peak density in the proposed **SBSP** beam.

At the heart of your microwave oven, the magnetron is a simple device making it especially suitable for the proposed **SBSP** beam at a fixed frequency. The microwave oven in my kitchen takes 1,800 watts of energy and concentrates it upon my cup of coffee. Yet, it still takes a full minute to warm it to a temperature that still will not burn my mouth. Our scientist's have shown that we can beam energy through the atmosphere using low-density microwaves without any adverse effects. It is time to tell our engineers to do it.

To those people who will never accept that **SBSP** microwave beams are harmless regardless of any amount of studies, yes, Apollo **DID** land men on the Moon, and no, the world did **NOT** end in 2012.

Bring it Home

The ground-receiving antenna is perhaps the best understood of the three major elements. Cell phone relay towers operate like FM radio stations, receiving and broadcasting in all directions. Microwave receivers come in a variety of shapes depending on the application. However, the incoming beam from a Solar Power Satellite is low intensity and unidirectional with very little leakage outside its footprint. This is the perfect application for a type of antenna called a rectifying antenna, or rectenna, which directly converts microwave energy into DC electricity. It consists of a multi-element phased array with a mesh reflector.

The basic rectenna element is a dipole antenna with a Schottky diode. The diode rectifies the AC current induced in the antenna by the microwave

beam and produces DC power. Schottky diodes have the lowest voltage drop and highest speed and therefore waste the least amount of power. This simple device is highly efficient at converting microwave energy to electricity, over 85%. The length of the dipole antenna itself depends on the specific frequency of the incoming beam but for those considered for **SBSP**, it is only a few centimeters long. Large arrays of many such elements are cheap and easily mass-produced.

A major advantage of this design is the ease that new rectenna's are constructed. A few hundred acres of land will supply the most remote area with electricity. The collection

Key Elements of a Rectenna
Dipole Antenna

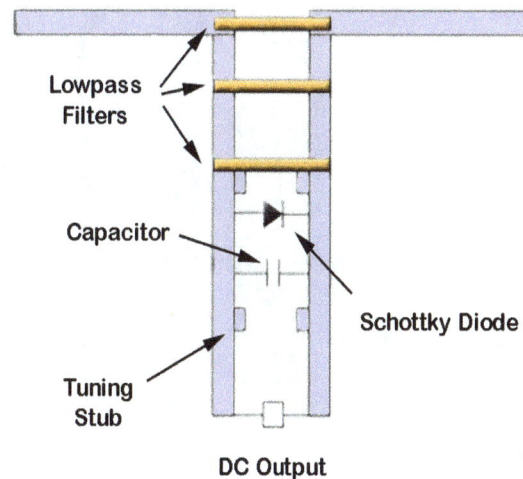

Lowpass Filters

Capacitor

Schottky Diode

Tuning Stub

DC Output

efficiency is so high that measurements taken beneath the rectenna array will pick up virtually zero microwave energy. Nothing's wasted. It is equivalent to the leakage around the door of your microwave oven while it heats your cup of coffee.

Eight Minutes to Orbit

Low Earth Orbit (LEO) realistically starts about a hundred miles up and extends out to 1,250 miles (160–2,000 km) but when I refer to LEO in this book, I generally mean the orbit of the International Space Station, about 200 miles. Not only must we lift every pound hundreds of miles straight up, we must also go from zero to 17,500 miles per hour to put it in orbit. The shuttle did this in about eight and a half minutes. Depending on how close you are to the equator, you can gain about a 1000 mph if you launch eastward with the spin of the Earth.

This is a rough comparison showing American, European, Indian, Russian, Japanese, and Chinese rockets and their payload capacity to LEO. Engineers designed these rockets to place payloads into orbit. There's no need to ask a machine designed for heavy lifting to shuttle cargo any further than LEO. Our engineers can design other machines better suited to take it from there. Besides, the ISS is the perfect base to assemble complex missions going on to geosynchronous orbit, Lagrange points, or the Moon.

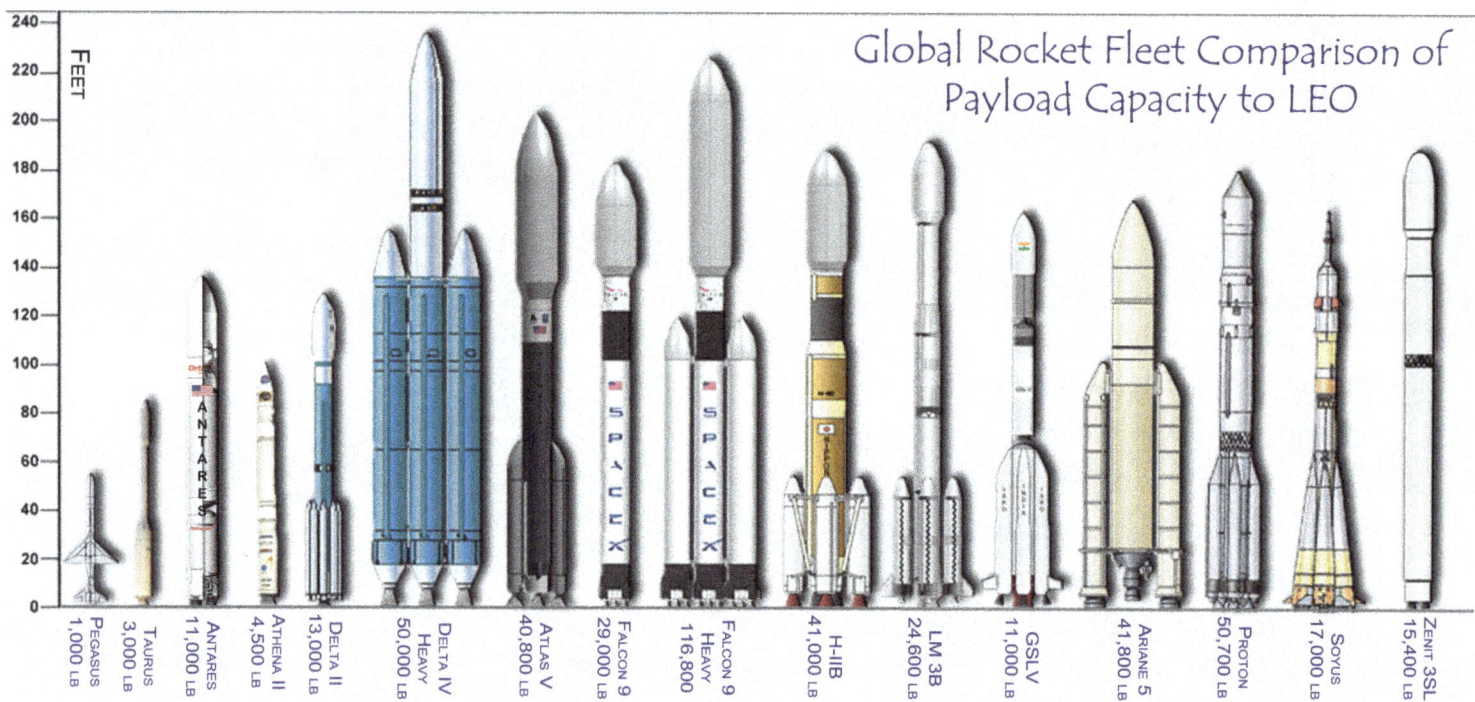

Global Rocket Fleet Comparison of Payload Capacity to LEO

Right now, the only way to acquire the necessary speed and altitude is with rockets. I'm sure that the future will bring other technology but for now, expendable rockets will do just fine. They will get our foot in the door, so to speak. Even without the Space Shuttle, the world has a nice little stable to choose from including some new kids on the block.

Starting with the smallest launcher, Orbital Sciences operates Pegasus, a winged vehicle capable of carrying small, unmanned payloads into LEO. A Stargazer L-1011 tucks Pegasus under a wing and carries it to 40,000 feet before letting it go, thereby allowing the rocket to avoid the densest part of the atmosphere. Even though it carries less than 1000 pounds, this is a great

resource for urgent deliveries. I can envision the Super Bowl commercial now… *Break a critical part? Need some extra oxygen? Burned out your life support system? **Never fear, Pegasus is here!** LEO delivery where you need it in twenty-four hours or less or it's **free**!*

Orbital also has Taurus, a four stage, solid fuel rocket able to carry a payload of around 3000 lbs into LEO. Still under development, Taurus II is a two stage, liquid fuel vehicle designed to launch payloads up to 11,000 lbs into LEO. NASA awarded Orbital Commercial Orbital Transportation Services (COTS) contracts to carry out eight resupply missions to the International Space Station using Taurus II and its Cygnus spacecraft.

Update: Orbital scraped the name of Taurus II and renamed their COTS launcher Antares. As of this writing, the Antares has made its first delivery to the ISS under its resupply contract and is schedule for a second mission in May, 2014.

NASA also awarded COTS contracts to SpaceX. The Falcon 9 is a two-stage liquid fuel rocket capable of placing 23,000 lbs into LEO. Under development, the Falcon Heavy is three Falcon 9's strapped together capable of putting 117,000 lbs into LEO which would make it the most powerful booster on the planet. SpaceX is advertising the Heavy at $95 million, which if they can do it, would bring the cost of putting that pound into orbit down to about $800.

If we drive down the cost of transportation in space, we can do great things.
CEO and Chief Designer of SpaceX and Tesla Car Company
Elon Musk

Update: As of April 2014, the Falcon 9 has made nine successful launches including three COTS resupply missions to the ISS. Three more ISS supply missions are scheduled for 2014.

Boeing launched the first Delta rocket more than forty years ago. Since then, more than 300 other launches have followed. Only two Delta rockets, Delta II and Delta IV, are in active use today. The Delta II is a three-stage liquid fuel rocket with up to nine solid rocket strap-on boosters delivering 13,000 lbs to LEO. Its big brother, Delta IV, is a family of launch vehicles maximizing the use of common hardware. The Medium & Medium-Plus configuration uses a single Common Booster Core (CBC), while the Heavy uses three CBCs strapped together, increasing the total payload to 50,000 lbs.

The Atlas rocket manufactured by Lockheed Martin has also been around since the sixties. The Atlas V uses a Common Core Booster (CCB) with up to five strap-on solid rocket boosters. Similar to the Delta IV, the core is a single stage liquid fuel rocket. Lockheed is developing a Heavy version using three cores strapped together. Atlas V has launched 26 times to date with only a single partial failure, an upper stage rocket engine flamed out early.

The current configurations of both the Delta IV and Atlas V are the result of an Air Force project, Evolved Expendable Launch Vehicle (EELV) which began in the 1990s. Its goal was to make launches more affordable and reliable for military projects. NASA recently awarded a contract to human rate the Atlas V.[1]

Lockheed Martin is the manufacturer of

---

1 7/18/2011: NASA, ULA sign agreement to develop human-rating Atlas V rocket

Athena. Begun in 1993, Athena was retired from service in 2001, and resurrected in 2010. It is set to resume flights in 2012. In September 2010, NASA added Athena to the Launch Services II contract. It is an all-solid fuel rocket capable of lifting 4,500 lbs into orbit.

The venerable Proton rocket was first launched in 1965 and is still in use today. It is one of the most successful heavy boosters in the history of spaceflight with a launch capacity to LEO of about 46,000 lbs. The other esteemed Russian rocket is the Soyuz, a three-stage liquid fuel vehicle with a 17,000 lbs payload, which according to the European Space Agency, is the most reliable launcher in the world. With the shuttle retired, the U.S. is dependent upon the Soyuz to ferry crews to and from the ISS.

A Ukrainian company and Boeing build the Zenit-3SL with a payload capacity of over 15,000 lbs. First flown in 1999, it has launched 30 times with two failures. It launches from the Ocean Odyssey platform anchored on the equator in the Pacific Ocean (154°W longitude) about 370 kilometers east of Kiritimati (formally Christmas Island). One of the two failures was a spectacular explosion at liftoff that decimated the platform.

In 1990, India initiated the Geosynchronous Satellite Launch Vehicle (GSLV) project. The GSLV is a three-stage launch vehicle with a solid fuel first stage, liquid fuel second and third stage, capable of delivering about 11,000 lbs into LEO.

Ariane 5 (European Space Agency) made its first successful launch on October 30, 1997. It is a two-stage liquid fuel rocket with solid fuel strap on boosters bringing its payload capacity up to 43,000 lb. The Ariane 5 launches the Automated Transfer Vehicle, a resupply spacecraft for the International Space Station.

The Long March 3B is the most powerful member of the Chinese rocket family. In 2003, China joined Russia and the U.S. as the only nations to have launched people into orbit. The LM 3B is capable of putting 24,000 lbs into orbit.

The H-IIB launch vehicle (Japanese

H-IIB

Aerospace Exploration Agency) is a two-stage liquid fuel with four strap-on solid fuel boosters. On January 22, 2011, the second H-IIB launched from the Tanegashima Space Center carrying cargo to the International Space Station.

The U.S. Air Force awarded a contract to Lockheed Martin in early December, 2011, to develop a Reusable Booster System (RBS). The RBS Pathfinder is a winged, rocket-powered flight vehicle that will demonstrate the fly back capabilities needed for the operational RBS.

For the RBS Pathfinder program, Lockheed Martin has also entered into an agreement with the New Mexico Spaceport Authority to conduct flight test operations from Spaceport America, the nation's first purpose-built commercial spaceport, located in southern New Mexico.

Update: The Pathfinder program was shut down in October 2012 due to funding cuts.

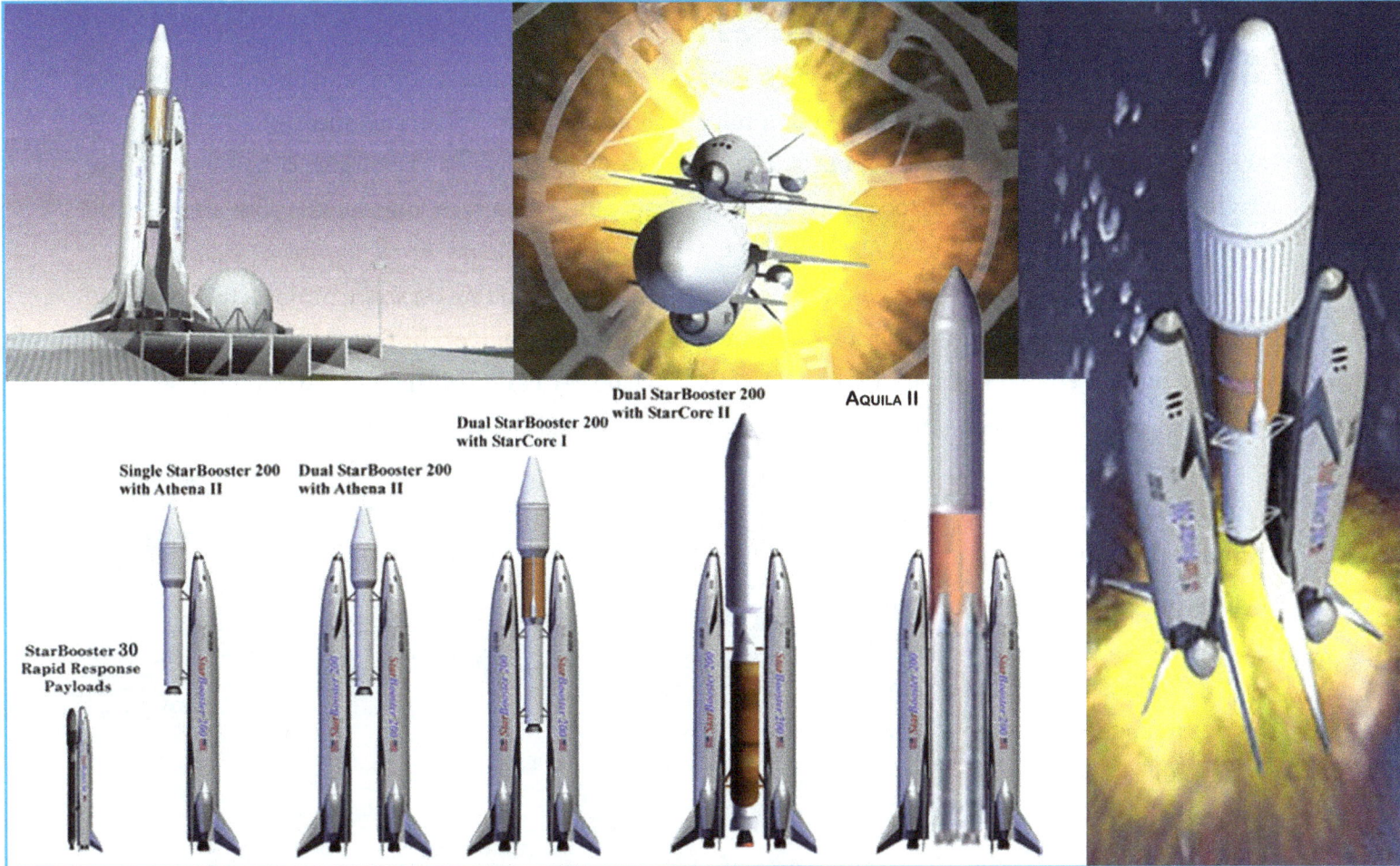

StarBooster 30
Rapid Response
Payloads

Single StarBooster 200
with Athena II

Dual StarBooster 200
with Athena II

Dual StarBooster 200
with StarCore I

Dual StarBooster 200
with StarCore II

AQUILA II

Buzz Aldrin

Astronaut Buzz Aldrin wore the second set of boots to touch the surface of our Moon. On July 20, 1969, he followed Neil Armstrong down Apollo 11's ladder and into the history books. The years since the glory days of Apollo have been hard on the explorers who actually walked on the Moon. They have endured the complete dismantling of the Saturn V and the painful lessons of the designed-by-committee Space Shuttle. As much as I admire the Space Shuttle, I must admit it had major problems. Not only was it our man-rated LEO launcher, but our heavy lifting cargo carrier, our orbiting laboratory, and our construction shack. A machine that tries to do so much does none of them well.

"Changing the way rockets are designed would pave the way to transporting more people into space. It could lead to a next generation type of space shuttle carrying up to 100 people."
Buzz Aldrin – Founder, Starcraft Boosters, Inc.

In 1996, Buzz founded Starcraft Boosters Inc. and began developing a family of reusable rocket boosters that applied the lessons learned during the shuttle program with the best of the expendable program. He felt there was a better way.

The StarBooster propulsion system is a two-stage liquid fuel core similar to the Atlas V, Delta IV or Zenit. What's different is how the vehicle is constructed. It is essentially a standard aircraft airframe with wings and fuel tanks, no cockpit or any of the weight associated with having a pilot aboard. After separating from the upper stage and payload, the unmanned StarBooster robotically flies the spent booster back to the launch site for refurbishment. Designed for quick removal, a module at the base contains the rocket engines. Change out the engine module, refuel, and your airship is ready to launch another payload in a matter of hours, not days.

In 2002, Starcraft introduced the Aquila, the Latin word for "eagle," named after the Aquila constellation on the east side of the Milky Way that resembles an eagle. The Aquila I carries a manned crew module named Altair (the largest star in the Aquila constellation). The Aquila II combines Starboosters with shuttle technology to lift heavy payloads. Aquila II would have the capability to accomplish large space missions, like building a gas station on the Moon, at a lower cost and risk, since fewer flights and less assembly in LEO would be required.

Seventy-two agencies around the world actively participate in space in some way. GPS has gone global, remote sensing is big and getting bigger, and having your very own satellite is more than just a status symbol, it is a necessity. If we limit the question to how many agencies are capable of launching payloads to LEO, the list shrinks to just ten. Only two are capable of launching people, Russia and China. America relinquished that capability when we retired the shuttle without a replacement waiting in the

wings.

There is one agency not listed that has the budget to go to space, the U.S. Military. As a veteran myself, I've never had a problem spending money on our military, but it has gotten a little out of hand. For a number of years the U.S. Military spending on space projects has exceeded the NASA budget. In fact, military projects take up nearly half of all spending worldwide on space assets and basic research.

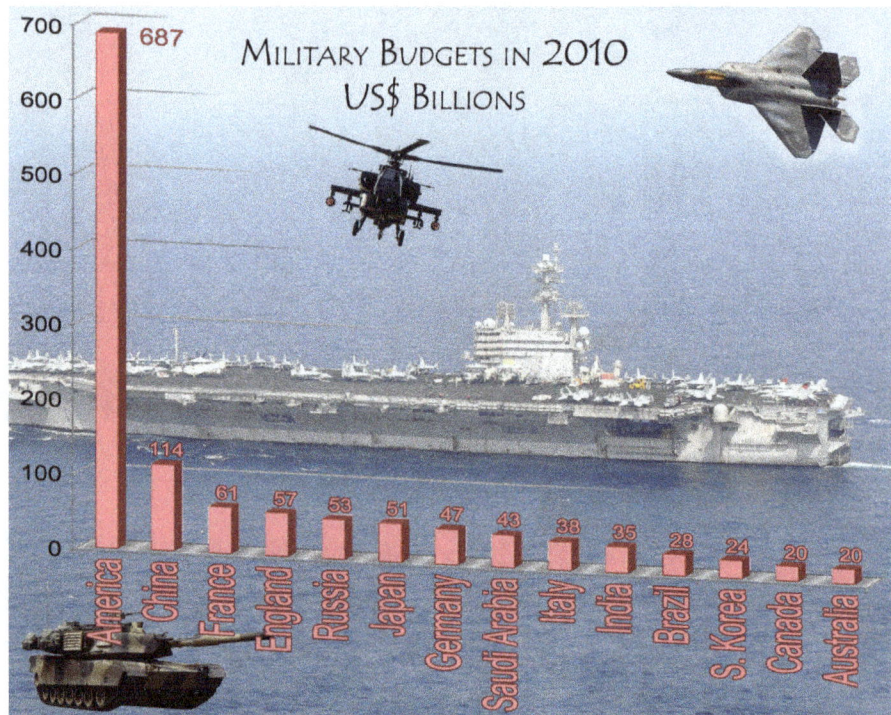

America is by far the biggest spender on defense-related space programs,[2] probably because we are the country most dependent on such systems. In their own words, the U.S. Military must dominate space in order to protect national interests and investments.[3] It is another Catch-22 situation; we cannot insure that another country cannot seize space without going there and doing it first.

2 9/1/2011: Space Security 2011
   http://www.spacesecurity.org/space.security.2011.revised.pdf
3 1997: Vision for 2020, US Space Command

Mass Driver

Colonizing space will require a lot of stuff, iron to build space stations, titanium to build spaceships, oxygen for us to breathe, and many other resources. Lifting all this up from the surface of the earth on rockets is simply not feasible. Thus, we will need to find these resources somewhere else. You need look no further than the moon. It has all the natural resources we need to colonize space but getting the metals and other material off the Moon and into cislunar space is a huge challenge. I don't believe we can rely on chemical rockets to get the job done, not because they can't lift it, they can, but because of the huge volume of material to be lifted. We have better things to do with the hydrogen and oxygen harvested on the Moon than burn it hoisting millions of tons of iron, aluminum, and titanium into lunar orbit.

Then the question remains, how do we get all this stuff into orbit? If rockets are not feasible, what is?

An idea emerged over a century ago called a Mass Driver. The first Mass Driver described in print was in the 1897 science fiction novel *A Trip to Venus* by John Munro. He called it an electric gun. It was his imaginative method of launching vehicles into outer space from the Earth's surface. Munro describes the electric gun as a series of coils energized in a timed sequence to provide the force necessary to get the spaceship into orbit.

Many SiFi authors have used these fictional devices in various ways, Arthur C. Clarke, Harry Harrison, James P. Hogan and Alastair Reynolds to name a few. By far my favorite SiFi mass driver is in Robert A. Heinlein's classic novel; *The Moon is a Harsh Mistress*. His plot has rebelling lunar colonists using a kilometers-long Mass Driver to bombard Earth and gain their freedom.

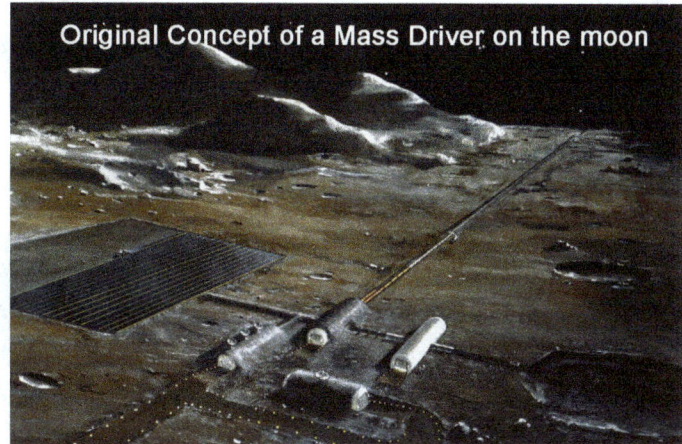

Original Concept of a Mass Driver on the moon

Putting a Mass Driver on the moon just makes sense. The moon has only 1/6th the gravity of Earth and with no atmosphere to slow things down, a Mass Driver could conceivably deliver payloads of ore or processed resources to a lunar orbit quickly, economically and in the quantities we need. Up until now, Mass Drivers have all been just another invention of science fiction, not of science fact, but that is changing very quickly.

In 1974, Professor Gerald O'Neill proposed Magnetic Levitation, or Maglev, to slingshot material off the Moon. He called his idea a Mass Driver. Professor O'Neill envisioned a long series of electromagnetic coils accelerating a stream of material into lunar orbit where it could be caught using a specialized catcher ship. This is essentially a coil gun on steroids, an ambitious project, to say the least, and impossible using 1974 technology.

However, realizing Professor O'Neill's vision is closer today than ever before. Chinese Maglev trains began commercial operation in 2003 running between the city of Shanghai and Pudong International Airport, twenty miles in

Six tugs nudge the GERALD R FORD toward the fitting out berth at Newport News Shipbuilding. (Huntington Ingalls Industries photo by Chris Oxley)

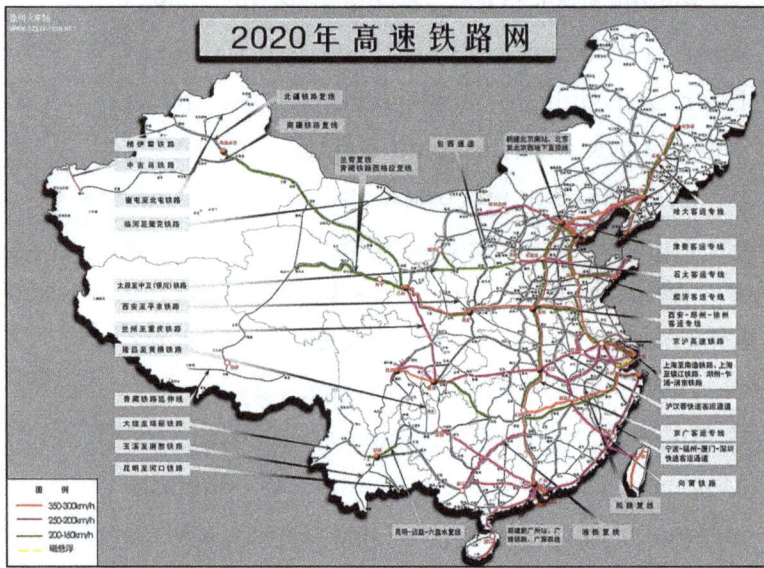

2020年高速铁路网

China's High-Speed Railway Network Map by 2020

seven minutes, averaging over 150 mph.[4]

In December 2010, another Chinese train broke the world record for fastest commercial trains, reaching 298.9 mph (481.1 kph).[5] According to Xinhua news agency, the new-generation CRH380 moved as fast as a low-cruising jet-plane during a trial run on what will become the country's newest rail line between Beijing and Shanghai.

4 How Maglev Trains Work
5 12/3/2010: Chinese train sets speed record, CNN Wire Staff

Ford Class Aircraft Carriers

As fast as the Maglev trains are, there's something faster. For over a decade, the U.S. Navy has been developing the technology to go electric. Let's start with the flagship of the U.S. Navy, the aircraft carrier. It is a fact that over the years, Navy airplanes have gotten heavier. Super Hornets tip the scale at around 24 tons, 20% more than the stocky F-4 Phantoms that flew over Vietnam from carrier decks. The Advanced Hawkeye is 2,500 pounds heavier than its predecessors. The Navy needed something better to get them off the deck than what they had. Their answer is the ElectroMagnetic Aircraft Launch System (EMALS). This all-electric catapult will replace the steam catapults currently used on aircraft carriers.

Northrop Grumman Shipbuilding is currently working on two of the US Navy's new Gerald R. Ford-class carriers in a shipyard on the Virginia coastline. Started in 2008, the first of these Supercarriers is already in the water undergoing testing and is scheduled to be commissioned

EMALS under development and installation.

CVN-78 Navy Photo

electromagnetic catapults. Why do I care? I care because electromagnetic's is one of the cornerstone technologies needed to colonize space. They are Mass Drivers.

There are many practical reasons why the Navy has developed the new electric catapults. The old steam catapults used about 1,350 psi of steam generated in the ships nuclear reactor to launch an aircraft. Steam catapults were mechanical nightmares consisting of a complicated web of hydraulics and associated high-pressure pumps, motors, and control systems. The result was a large, heavy, maintenance-intensive and dangerous system that operated without any feedback control. It inflicted sudden shocks to the aircraft it was launching which shortened their lifespan.

All that is changing with the new all electric Navy. Taking the place of the steam catapult will be the Electro-Magnetic Aircraft Launch System, EMALS or simply EM catapult. It uses the electromagnetic forces generated from extremely high currents to accelerate the shuttle holding the aircraft. The force thus generated provides a much smoother launch and 30% more energy to cope with today's heavier planes. The EM catapult also has far lower space and

in 2016 at the cost of over $13 billion. These new ships have three times the electrical power generating capacity compared to Nimitz-class carriers. The USS Ford will have four 26-megawatt generators bringing a total of 104 megawatts to the ship.

Built on the basic footprint of the earlier Nimitz-class carriers, that is where the similarity ends. The very heart of any aircraft carrier is obviously its ability to launch planes. Aircraft carriers are floating airports with a crew of thousands. In these new ships, the Navy has abandoned the old steam catapults and gone all electric. The ship is designed around powerful maintenance requirements. Ancillary benefits include the ability to embed diagnostic systems and increasing the ease of maintenance with fewer personnel. Is it just me or does this sound perfect for lunching something off the surface of the Moon?

Beginning in 2004, more than 1,500 launches on a full-scale prototype fundamentally proved EMALS technology. Testing on the first generation, aircraft carrier design began in 2008, including full scale/full power functional tests of all future shipboard components.

Testing the new catapult with manned aircraft began in December 2010 with the launch of an F/A-18E Super Hornet fighter.[6] Late September 2011, an E-2D Advanced Hawkeye became the 96th aircraft tested on the new catapult.[7] It went flawlessly. The Navy's transition to electromagnetic flight decks is almost complete. The first Ford class aircraft carrier using EMALS is in the water undergoing sea trials right now and will be commissioned in 2015 with nine more to follow. The contract is worth billions for Northrop Grumman Newport News in Virginia.[8]

Compared with steam catapults, the EMALS technology is more powerful and fully capable of launching heavy aircraft at higher speeds. In addition, it has several major advantages over current steam catapults; it lowers overall operating costs, reduces maintenance, and greatly reduces the launch stress on carrier-based aircraft. This also translates into less stress on the ship and increased safety for the flight crew. The Navy's going electromagnetic and it is about time.

Naval Railguns

However, as exciting as the Navy's new catapult is, something else they have been developing is even more electrifying. The Electromagnetic Railgun is the Navy's replacement for gunpowder-style cannons and uses the same electromagnetic technology as the catapult but instead of accelerating a 24-ton fighter from zero to 150 mph in a few seconds, it flings a 23-pound (10 kg) projectile from zero to 5500 mph (2.45 km/sec) in a few milliseconds. In December 2010, a prototype of the Navy's newest standoff canon passed another major milestone when the first shot of the day generated 33 mega joules (MJ) of kinetic energy out of the barrel, three times the current world record for muzzle energy.[9]

To fire a round, a million amp electrical current is sent through the conducting rails generating an enormous electromagnetic field. The force generated by the electromagnetic field is called the Lorentz Force. It wants to push the rails apart but when that is not possible, it pushes the payload instead. The Railgun uses what is known as a sabot to encase the aerodynamic round in metal designed to withstand the tremendous forces and temperatures involved in the process. The sabot falls away after launch but not before accelerating the projectile inside to extreme high-velocity.

A good way to get a handle on the magnitude of this accomplishment is by comparing it with something we all should know about, a car. One mega joule (MJ) is roughly equal to the energy generated by a one-ton vehicle moving at 100 mph. A one-ton car isn't very big. Picture thirty-three Smart Cars (1,700 lbs curb weight) barreling

6 Naval Air Systems Command – Aircraft and Weapons – Electromagnetic Aircraft Launch System
7 9/28/2011: Navy pleased with new electromagnetic catapult, Kirk Moore
8 4/20/2011: Navy Ford (CVN-78) Class Aircraft Carrier Program: Background and Issues for Congress - RS20643
9 12/10/2010: Navy's Mach 8 Railgun Obliterates Record

down the freeway at 100 mph, bumper to bumper, their little engines screaming like banshees, an NFL lineman at the wheel of each car. Now, put all that crazy energy into something the size of a football… Say hello to the Navy's little friend!

Firing an average of twenty shots per week, by late October 2011 they had fired the Railgun 999 more times setting a materials testing milestone in the weapon's development. The ultimate goal is to fire the gun every few seconds at 64 mega joules, making it capable of sending a projectile two-hundred miles in six minutes.[10] That's ten times farther than the Navy's chemical powered guns can fire, and at a much faster rate.

Think about this, if it can deliver a shell two-hundred miles downrange, it can also engage targets in LEO. A projectile fired straight up will penetrate space to an altitude of about 300 miles where any collision with an orbiting object will have closing velocities of at least 17,500 mph. The kinetic energy of such a collision is enormous. Of course, what goes up must come down so the Navy had better keep track of the pieces.

The electromagnetic Railgun is a major weapons development program that will make explosive-based cannons a thing of the past. For decades, the Navy has been working towards arming warships with battle-ready electromagnetic Railguns. This isn't surprising considering that such a weapon has the potential to intercept missiles with an unparalleled combination of long-range accuracy and extreme velocity. Without much fanfare, in 2012 the Navy took possession of the first two prototype Railguns from General Atomics and BAE Systems.

During testing of the General Atomics Blitzer Railgun, rounds fired from the gun blasted right through a 1/8-inch thick steel plate located 100

10 10/31/2011: U.S. Naval Research Laboratory Materials Testing Facility Press Release 145-11r: Daniel Parry

Railgun – a 21st-century weapon

In the opinion of the U.S. military, electromagnetic weapons have the potential to replace conventional artillery in the near future

(!) *The most powerful railgun in the world was designed at the U.S. naval research laboratory in Dahlgren, Virginia. The energy of its rounds is 33 megajoules. Projectile velocity is five times the speed of sound and its firing range can reach 370 km*

Railgun device

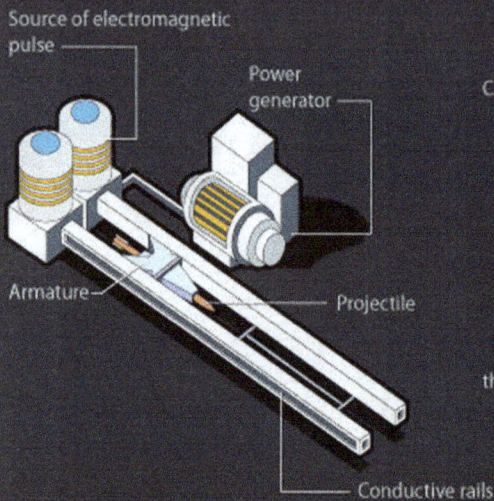

Source of electromagnetic pulse

Power generator

Armature

Projectile

Conductive rails

Interaction of magnetic fields

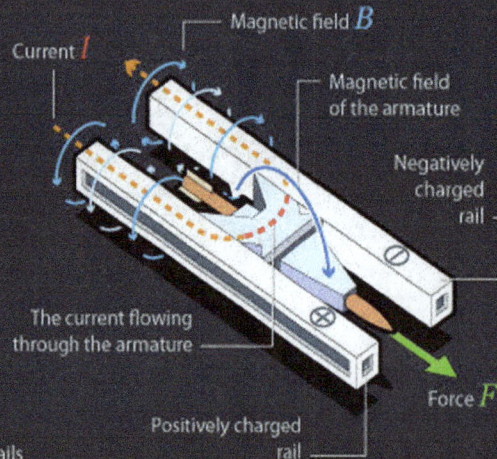

Current I

Magnetic field B

Magnetic field of the armature

Negatively charged rail

The current flowing through the armature

Positively charged rail

Force F

The principle behind Lorentz force

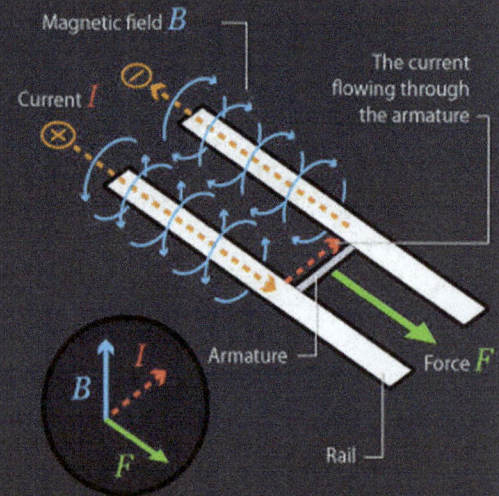

Magnetic field B

Current I

The current flowing through the armature

Armature

Force F

Rail

B I F

The railgun uses electromagnetic force (Lorentz force) to propel an electrically conductive projectile that is initially part of a chain. Current I, flowing through the rails, generates magnetic field B in the rails and armature. As a result, under the action of force F, the armature is pushed out of the magnetic field of the rails and the projectile accelerates

meters downrange at Mach 5 (about 4,000 mph) and continued to travel more than four miles at zero elevation. The BAE Systems Railgun did even better. It achieved a muzzle velocity of 5600 mph on a 23 lb payload. The escape velocity of the moon is only 5300 mph. Mission accomplished! Right?

Why is this exciting? Why should I, an advocate of civilian space colonization care about the newest gun in the Navy arsenal? The muzzle velocity of their current generation Railgun is 5500 mph (or about a mile per second) and the escape velocity of the Moon is 5325 mph. The Navy has developed the perfect machine for getting massive quantities off the surface of the Moon, twenty pounds at a time. Moreover,

it is not kilometers long but the size of a semi trailer. All we need to do is engineer a way to get it on the Moon, power it up and supply it with an unending stream of thick iron shells filled with almost anything that doesn't mind a 15,000 gee kick in the ass. This isn't as farfetched as it was just a few years ago. Our scientists know the science and our engineers have been developing hardware for over thirty years.

This technology operated on the Moon has many advantages over its Earth-bound cousins. The Navy designed their rail gun to operate onboard a warship, the very definition of sea level. The projectiles bull their way forward superheating air molecules, shedding energy, and losing velocity. Lunar vacuum eliminates the

Electromagnetic Railgun

aerodynamic effects associated with accelerating an object at the bottom of our thick atmosphere from zero to 5500 mph in less than a second. I imagine the sonic boom that occurs when the Railgun fires is tremendous. On the Moon, complete silence.

Another huge advantage is the incredible cold on the Moon when the Sun isn't shinning. The low temperature of a lunar night varies from -294°F to -387°F.[11] Under normal atmospheric pressure, nitrogen exists as a liquid between the temperatures of -346°F and -320°F (63K and 77.2K). Below -346°F, nitrogen freezes and becomes a solid. Above -320°F, nitrogen boils and becomes a gas. The machines we send to the Moon to do our bidding can take advantage of this large temperature difference.

The Navy intends to fire their Railgun every few seconds[12] but even if our Mass Driver fired once a minute, at twenty pounds per shot a single unit would deliver fourteen tons of iron and other materials to lunar orbit in a **single day**. This is doable. It is only a matter of engineering and the political will to pull it off.

Even at this early stage, the Railgun is already capable of launching a 23 pound payload off the surface of the moon. We have our Mass Driver but getting it operational on the moon along with all the support needed to supply it with payloads.

11 9/27/2000: The Artemis Project, Lunar Surface Temperatures, Marvin Ostrega

12 10/22/2009: General Atomics Blitzer Electromagnetic Railgun Completes Successful Fisrt Firing

More Military Technology

Several recent Defense Advanced Research Projects Agency (DARPA) space missions have raised concerns about militarizing space. Government officials feed the fire when they refuse to reveal the exact purpose of Boeing's X-37B Teleoperated Space Plane and the Falcon HTV-2 Advanced Hypersonic Weapon.[13] The first X-37B mission started atop an Atlas V rocket. It launched from Cape Canaveral, Florida, on April 22, 2010, and landed at Vandenberg AFB on 3 December 2010. The second launch was on 5 March 2011 and the ship is still in orbit as of this writing. The Air Force will only say these are technology test vehicles. They wouldn't even provide the orbital data necessary to calculate the paths. This lack of information spawned many speculations most of them involve spying on China.[14]

Tiangong is China's new spacelab, in orbit since September 29, 2011. Its first crew is scheduled to arrive sometime in 2012. Would it be a great surprise to find out we are spying on it? But there are many experts, Brian Weeden among the most vocal, that say the X-37B's orbit makes it nearly impossible to spy on the Chinese space station. It is far more likely that the X-37B is just the latest asset in our fleet of spy satellites. It adds two capabilities all the Keyhole

spy satellites lack, a cargo hold about the size of a Ford truck bed and the ability to deorbit and come home.

There's no doubt about the mission of the Advanced Hypersonic Weapon. Called the Falcon, it is designed to strike targets anywhere in the world in under an hour.[15] They are shooting

X-37B Orbital Test Vehicle

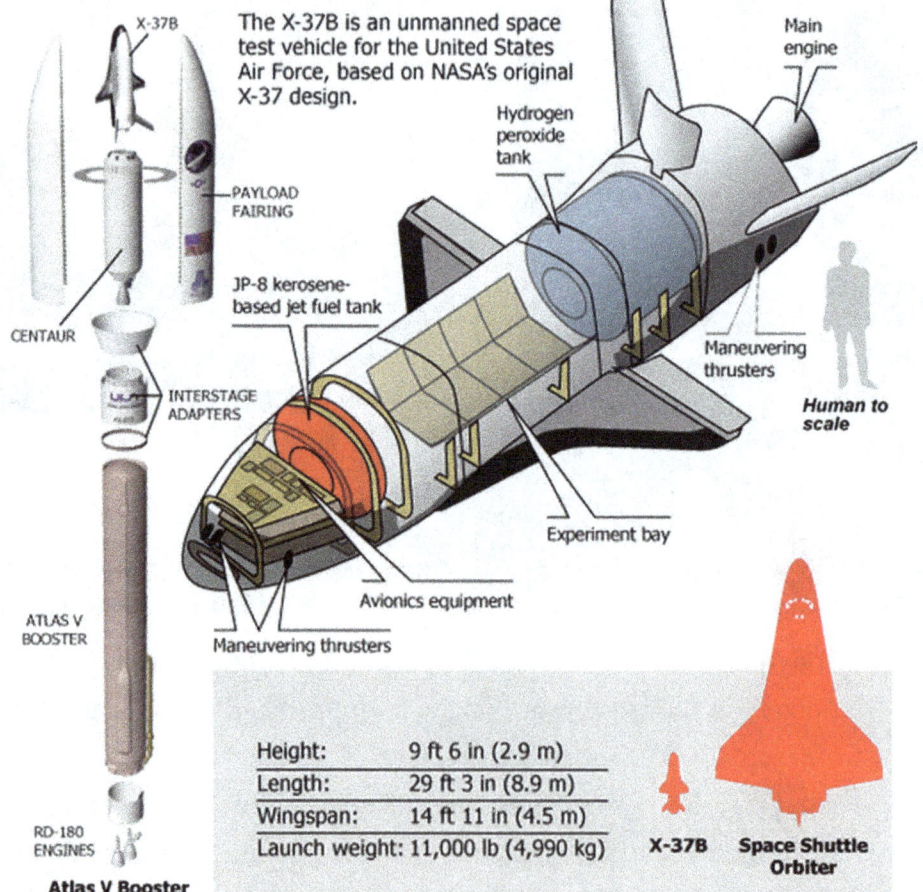

The X-37B is an unmanned space test vehicle for the United States Air Force, based on NASA's original X-37 design.

X-37B
PAYLOAD FAIRING
CENTAUR
INTERSTAGE ADAPTERS
ATLAS V BOOSTER
RD-180 ENGINES
Atlas V Booster

Main engine
Hydrogen peroxide tank
JP-8 kerosene-based jet fuel tank
Maneuvering thrusters
Human to scale
Experiment bay
Avionics equipment
Maneuvering thrusters

Height:	9 ft 6 in (2.9 m)
Length:	29 ft 3 in (8.9 m)
Wingspan:	14 ft 11 in (4.5 m)
Launch weight:	11,000 lb (4,990 kg)

X-37B **Space Shuttle Orbiter**

SOURCE: NASA, United Launch Alliance   Graphic by Karl Tate

for a mind-boggling 13,000 mph. That's within striking distance of orbital speed, 17,500 mph. Unlike a ballistic missile, a Falcon can maneuver and avoid flying along a predictable path. The first flight was 22 April 2010, the second was 11 August 2011. Unfortunately, they both failed about nine minutes after separation from their launchers. Don't let that worry you. Engineers

13 Falcon HTV-2 - DARPA
14 1/5/2012: X-37B spaceplane 'spying on China'
15 11/19/2011: Army tests its new hypersonic weapon, AP, Honolulu

HOW THE ADVANCED HYPERSONIC WEAPON WORKS

FACTFILE
- Length: 12ft
- Weight: 900kg
- Materials: Made of experimental carbon composite that can withstand up to 3,500f (2000C) which it will experience in flight – hot enough to melt steel.
- Range: Anywhere in the world in less than 60 minutes
- Cost: £189m so far

Payload: Potentially anything up to 12,000lbs, including a nuclear bomb or a missile.

Missile

London to ...Sydney
Jumbo Jet: 23 hours
Falcon HTV-2: 1 hour

Controls: Unmanned, but sensors will take millions of temperature, navigation and speed measurements.

Thrust: Turbo jets ignite during flight for cruising, then Scramjets kick in at the hypersonic phase. They will take it to 13,000mph or Mach 22.

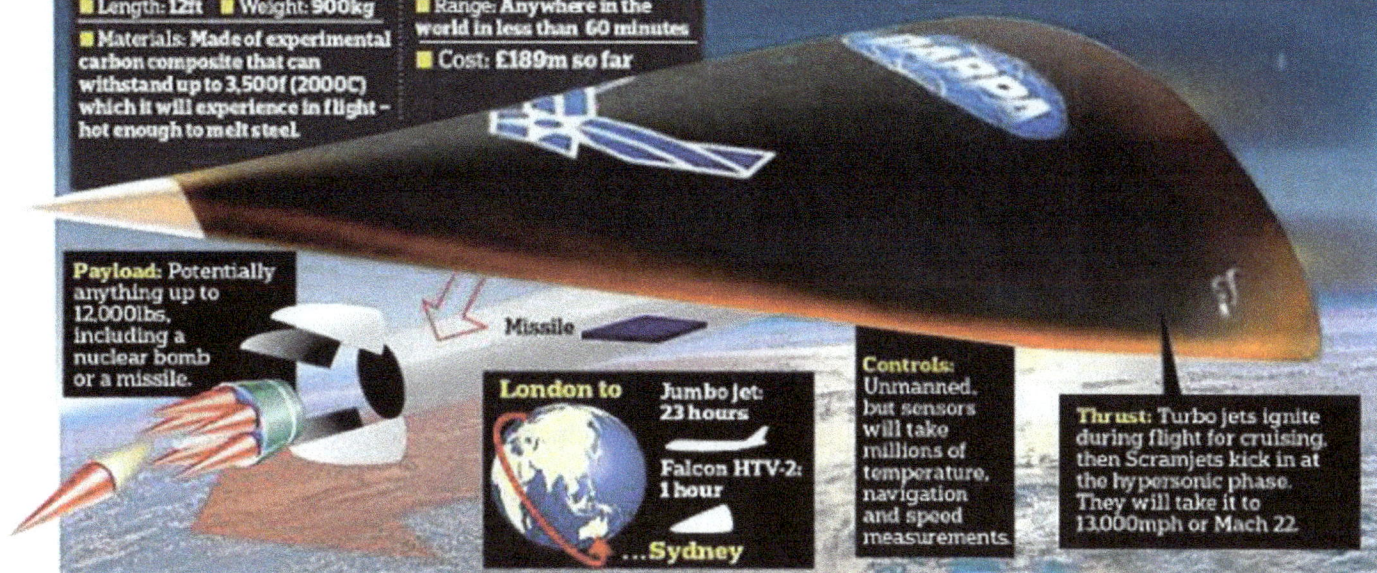

learn more from failures than they ever do from successes.

These DARPA projects are among a profusion of new technologies that are bringing space under control of the U.S. Military. What we see is just the tip of a robust space-weapons program buried within the Pentagon's huge budget. Black projects are even harder to find.

In 2007, China demonstrated the ability to destroy stuff in orbit, sending an anti-satellite (ASAT) device against one of their weather satellites. A year later, the U.S. Military likewise shot down one of their own spy satellites with a sea-based missile. India is also developing an anti-satellite program combining lasers and kinetic kill technology. The problem, blowing things up in orbit creates huge swarms of space debris that threaten other spacecraft. All it takes is a few grams impacting at 30,000 mph to ruin your day.

Besides the obvious ground launched space weapons, some argue that orbital space is already militarized. So much of the technology goes far beyond dedicated military objectives and serves both civilian and military masters. Many military navigational and targeting systems depend on Global Positioning System (GPS) satellites that also guide civilian cell phones and other location devices from golf carts to interstate trucking all around the world. A launcher capable of delivering a satellite into orbit can deliver a bomb anywhere on Earth's surface. A satellite capable of making a million internet connections can connect an army or guide a drone in for a strike. A weather satellite can easily provide invaluable battlefield data. Imagine what Napoleon would have given for Google Earth!

The military has already designated **Space Based Solar Power** as an issue of national security. It doesn't take a genius to conclude that they are doing something about it. The host of Army, Navy, Air Force and DARPA projects prove that. I can only hope they have projects to harvest water on the Moon, smelt metals from lunar ore, and build the first Powersat.

Superconductors

A major obstacle to rapid fire is the heating of the rails as the projectile is accelerated. High force equal high stress which creates high heat. Even though the Navy's design uses superconductors to handle the huge current, the rails require re-cooling after every shot. Engineers can design the lunar systems to take advantage of the cold to cool the Railgun (or should I say Mass Driver) after each shot.

Superconductors are materials that offer zero resistance to the flow of electricity. In other words, rails made of a superconductor will not get hot as more and more electricity passes through it. The phenomenon was first observed in 1911 by Dutch physicist Heike Kamerlingh Onnes after he had cooled mercury to 4° Kelvin (-452°F, -269°C), the temperature of liquid helium. To induce superconductivity in pure mercury, it was necessary for Onnes to come within 4 degrees of Absolute Zero, the coldest temperature that is theoretically attainable. By experimentation, he discovered other materials would also exhibit superconductivity, each at its own point known as the transition temperature, or Tc. His research won him a Nobel Prize in 1913.

Twenty years later, Walter Meissner and Robert Ochsenfeld discovered that superconducting materials would energetically repel a magnetic field. This phenomenon is officially named diamagnetism but is often referred to as the Meissner effect.

In the decades that followed, scientists discovered other superconducting materials such as niobium-nitride, vanadium-silicon, and an alloy of niobium and titanium, to name a few, but there was a problem. It seemed every material had a different hypothesis to account for its superconductivity. None provided a single unifying theory that spanned all the compounds. What explained one superconductor, unraveled with the next.

To make matters worse, in the 1980's scientists found a second type of material exhibiting superconductivity. Alex Müller and Georg Bednorz, working at the IBM Research Laboratory in Rüschlikon, Switzerland, created a brittle copper-oxide ceramic compound that superconducted at the highest temperature then known: 30°K (-405°F, -243°C). These became known as Type 2 superconductors or cuprates. A unified theory of superconduction seems even further away.

Research into Type 2 materials continued as more and better superconductors were devised, each striving to push the transition temperature ever higher. The current record holder is $HgBa_2Ca_2Cu_3O_x$ with a Tc of 135°K (-216°F, -138°C), a temperature easily maintained using liquid nitrogen or the clever use of the coldness inherent in space.

The world's first high temperature superconducting power transmission lines entered commercial service in 2008.[16] The development of new superconductors will continue to drive our ability to take advantage of electromagnetism into the future. Who knows, maybe someday we can hold the power of a Railgun in the palm of your hand. That should make the NRA drool.

I look at our huge military budget and hope that more of the technology they research has civilian applications and that someday soon, they will be ready to share it with the rest of us.

---

16 7/1/2008: Superconductors Enter Commercial Utility Service, ieee spectrum

Mining the Moon

Low Earth Orbit is halfway to anywhere. This is referring not to a place, but to a speed. To get into LEO, we need to not only lift our payload at least 100 miles straight up, but also accelerate it to 17,500 miles per hour (mph) ground speed. Technically, Earth's escape velocity is 25,000 mph so you only need to add another 7,500 mph and you can, in theory, go anywhere in the solar system. In practice, it's more complicated than that, but the point is, the major effort is in getting to LEO. From there, you're halfway to anywhere.

Engineers design expendable launchers specifically to perform this heavy lifting. To ask more would be like asking Southwest Airlines for a ride to the local market in one of their 737s. Launchers exist for the sole purpose of lifting stuff out of Earth's atmosphere and making it go fast. Other machines better suited for work in orbital space can take it from there. It doesn't take the raw power of a shuttle main engine to move about cislunar space, the region between LEO and the Moon. For instance, the Saturn V that launched the Apollo spacecraft to the Moon expended the entire first and second stages, as well as some of the fuel from the third stage, just to get off the ground and into orbit. From there, a burn of a few minutes from the small third stage was enough to send them on a 250,000-mile journey to the Moon.

Let's look at the thrust and specific impulse characteristics of a propulsion system. In simple terms, thrust is your engines horsepower and specific impulse is its gas mileage. Thrust is defined in Newton's Third Law, *for every action, there is an equal but opposite reaction.* Expendable launchers depend on this to bull their way through the atmosphere and into orbit. It is what they do.

Chemical rockets get their horsepower from burning through a large amount of fuel in a short time. That's not good for the gas mileage side of things but it does get you up the hill in a hurry. For example, a shuttle main engine at sea level has 420,000 pounds of thrust with a specific impulse of 360 but it burns through thousands of pounds of fuel every second. However, once we're in orbit, we don't need this level of power to get to where we are going. We can use something with a little less horsepower but long on gas mileage, a machine that maximizes the specific impulse. Enter the Variable Specific Impulse Magnetoplasma Rocket (VASIMR) the brainchild of Dr. Franklin Chang-Díaz.

> *"It's a plasma-based electric rocket engine, so it's different from conventional chemical rockets, which are propelled by the combustion of rocket fuel. VASIMR isn't based on chemical reactions. Instead, it uses plasma, which is a gas that's been heated to extremely high temperatures*

Newton's Law

$$Force = Mass \cdot Acceleration$$

$$f = ma$$

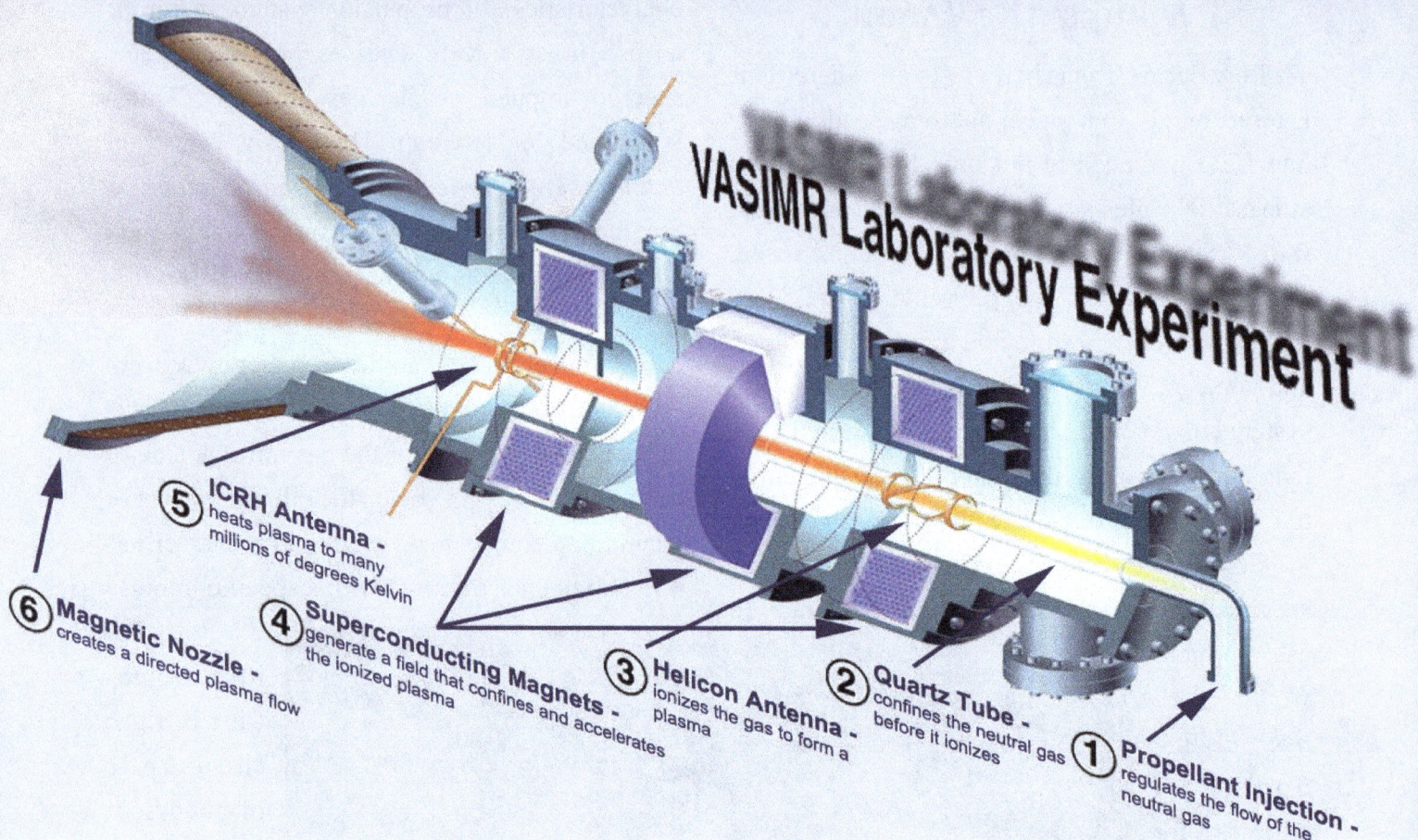

VASIMR Laboratory Experiment

⑤ **ICRH Antenna -** heats plasma to many millions of degrees Kelvin

⑥ **Magnetic Nozzle -** creates a directed plasma flow

④ **Superconducting Magnets -** generate a field that confines and accelerates the ionized plasma

③ **Helicon Antenna -** ionizes the gas to form a plasma

② **Quartz Tube -** confines the neutral gas before it ionizes

① **Propellant Injection -** regulates the flow of the neutral gas

approaching that of the Sun. Because it's so hot, the plasma can't be handled with any conventional materials. We have to use superconductors to generate electromagnetic fields to contain the plasma, form it into a jet, and guide it out the back of the rocket engine. VASIMR is meant for use in outer space—it won't replace chemical rockets for launching payloads into orbit."

Dr. Franklin Chang-Diaz

◇◇◇◇◇◇◇◇◇◇◇◇◇◇◇◇◇◇◇◇◇◇◇◇◇◇◇◇◇◇◇◇◇

His company, Ad Astra Rocket Company is completing ground testing on the VX-200 Thruster in preparation for taking it aboard the International Space Station for some on-the-job training.

Using much the same technology as a Railgun or the Super Collider, the VASIMR accelerates ionized gas to high speed using magnetic fields, then ejecting them at velocities much higher than a chemical rocket. Even this first-generation thruster achieves exhaust velocities above 30 miles per second. (The shuttle main engine is at just over two miles per second) The resulting thrust of a VASIMR is much lower than that of a shuttle engine and depends on accelerating small bits of plasma to high speed but doing it for months without stopping. Where chemical rockets exhaust thousands of pounds at a relatively slow speed in minutes, VASIMR exhausts a few pounds at very high speed in days.

In other words, the faster the exhaust, the more thrust is produced from the same amount of propellant but spread out over a much longer period. This technology represents the future of

translunar and interplanetary transportation. The VASIMRs superb efficiency compared to that of a conventional chemical rocket delivers larger payloads into lunar orbit for much less money. The VASIMR can thrust for weeks, months, even years at a time. A typical lunar cargo delivery mission will take over a hundred days to spiral out from LEO and arrive in lunar orbit, its VASIMR gently but relentlessly pushing it along without stopping. It may take longer but the cargo gets there just the same.

A VX-200 with a thrust of only a few pounds has a specific impulse of over 5000 seconds but consumes a tiny fraction of the fuel that a shuttle main engine does. What this means is that while the magnetoplasma thrusters cannot even lift themselves off the ground, once in orbit, they can move our payloads around cislunar space with ease, as long as you're not in a hurry.

This capability is something we are going to need if we are to solve our coming energy crisis and harvest sunlight in space. Utilizing lunar resources to construct humongous solar power satellites in geosynchronous orbit will require moving a great deal of unmanned cargo across cislunar space. Water processing facilities, metal refining, robotics, and the supplies to keep everything working goes to the Moon. The iron, aluminum, titanium and perhaps even solar cells come from the Moon. The old way relied solely on chemistry. The new way uses a fleet of electric powered vehicles to do the job much more efficiently.

Chemical engines still have their place. Even at $1/6^{th}$ that of Earth, the gravity field of the Moon is still too large for the first generation VASIMR to lift its own weight off it, let alone a payload.

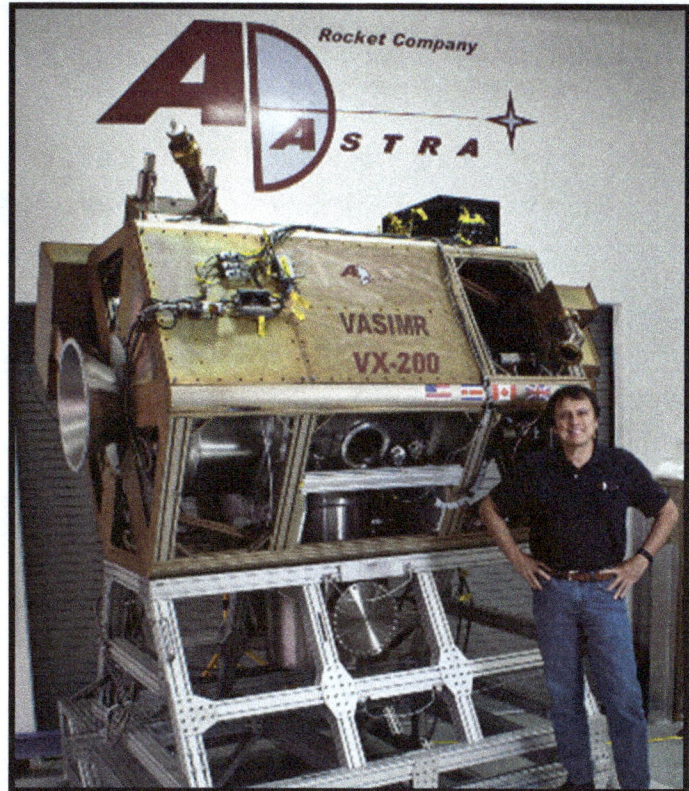

This will require another specialized machine, the Moon lander, powered by restartable liquid fuel engines. Unlike the expendable launchers, a lander will live out its life in the vacuum of space and thus, needs no aerodynamic considerations. A solid design will visit the Moon many times before it is retired.

Hydrogen is the fuel of choice for both the VASIMR and the chemical engines needed to land and take off from the Moon. In October 2010, NASA confirmed the existence of a billion gallons of water ice in the deep recesses of Cabeus Crater near the lunar South Pole. You can bet this will be the first thing we go after.

Using electrolysis and electricity harvested from the almost constant sunlight at the poles, we can split water into its constituent elements, oxygen and hydrogen. Granted, a billion gallons is not very much, about 1500 Olympic-sized swimming pools, but this is only one crater.

Once we have robotic prospectors poking about the Moon, there's no telling what we will find. Getting our fuel from the Moon instead of using expensive rockets to lift it off the Earth is a tremendous savings in both time and money. However, the Moon has much more to offer than just water.

Our ancestors undoubtedly considered everything in the sky to be supernatural at some time or the other. Gods and goddesses, heroes and villains, fallen kings and beloved fathers, all came to life in the sky above. The Moon played a dominant role in virtually all these philosophies, and well it should. Just like the Sun, the Moon is our constant companion.

In 1546, English playwright John Heywood made one of his most famous epigrams claiming: *"The Moon is made of a greene cheese,"*[17] greene meaning new, or unaged. I'm hearing sarcasm here, but one can't be sure. Other citations are clearer: *"You may as soon persuade some Country Peasants, that the Moon is made of Green Cheese (as we say) as that 'tis bigger than his Cart-wheel."*[18]

In 1609 and 1610, an Englishman named Thomas Harriot produced some of the earliest written observations of the Moon that we have. Using his telescope, he made detailed drawings of the lunar phases and even made maps of its topography.[19] For the next 350 years, others followed his example with better and bigger instruments.

In 1957, the Soviet Union launched Sputnik and drew the United States into the Space Race.

Some fifty robotic missions later,[20] America won when Apollo 11 touched down at Tranquility Base on July 20, 1969. A human walked on another heavenly body for the first time.

In total, twelve men walked on the surface of the Moon and brought back some 840 pounds of rock samples. The Apollo Program taught us a lot about what makes up the Moon and showed the world that we could actually go there. (No, they didn't find any dairy products.) When the politicians killed the Saturn V program, they killed our ability to send people to the Moon. It would be almost twenty years before we again sent robotic explorers to orbit the Moon. Men have yet to return.

On January 25, 1994, Clementine orbiter began scanning the lunar surface at a number of resolutions and wavelengths from UV to IR. The most relevant data it obtained was topography maps of surface deposits of titanium, iron, and thorium. Four years later, the Lunar Prospector extended the data using upgraded instruments. Moon Mineralogical Mapper and now Lunar Reconnaissance Orbiter have added to our knowledge.

Japan and China have recently sent orbiters to the Moon to further refine the resource data. We know about all there is to know about the surface of the Moon from orbit. It is apparent the Moon has many, if not all, the raw materials to build a space faring civilization. We just need to be clever enough to go get them.

Regolith is the layer of pulverized solid material covering the surface of the Moon. Constituents vary but as a rule of thumb, oxygen,

---

17 1562: The Proverbs and Epigrams of John Heywood
18 1638: Wilkins, New World 1
19 1995: The Galileo Project, Thomas Harriot

20 LRO, LCROSS, Moon Impact Probe, Chandrayaan-1, Kaguya, Chang'e 1 & 2, SMART-1, Lunar Prospector, Clementine, Pioneer, Luna, Lunokhod, Surveyor, Ranger, Zond, Hiten, Galileo, Explorer, Lunar Orbiter

silicon, iron, calcium, aluminum, magnesium, and titanium are present. We just need to engineer a way to separate them efficiently using robotic systems teleoperated from Earth.

For the last half century, our scientists and engineers have explored ways to harvest lunar resources taking advantage of the lunar environment. Getting electrical power from sunlight makes sense, it is free and we don't need to carry it with us. Besides electricity, solar furnaces are capable of producing very high temperatures. The world's largest solar furnace is the Font-Romeu-Odeillo-Via in France. It is capable of temperatures up to 6330° Fahrenheit. Without the atmosphere absorbing a good percentage of the sunlight, a furnace located on the Moon will be more efficient.

Molten Regolith Electrolysis

Simply heating the lunar regolith will release any gases trapped there including hydrogen, helium, nitrogen, and carbon. Further heating will produce sulfur, chlorine, argon, water, hydrogen sulfide, carbon monoxide, carbon dioxide, ammonia, and hydrogen cyanide.[21] Bring it to a boil and you will have molten metals. This is called Molten Regolith Electrolysis. It will extract gaseous oxygen and produce high quality metals and glass from regolith in a single-step electrolysis of the minerals.

Molten oxide electrolysis is an extreme form of molten salt electrolysis, a technology that has been supplying tonnage metal for over a century. Aluminum, magnesium, lithium, sodium, and the rare-earth metals are all produced in this manner.

Clementine Iron Map

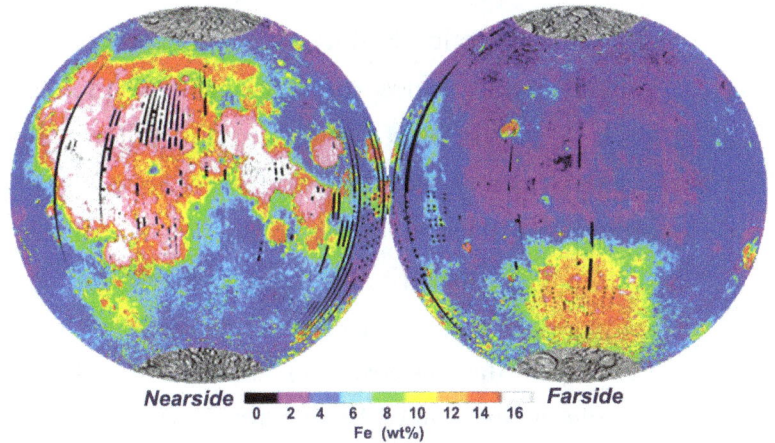

Nearside   0   2   4   6   8   10   12   14   16   *Farside*
Fe (wt%)

Clementine Titanium Map

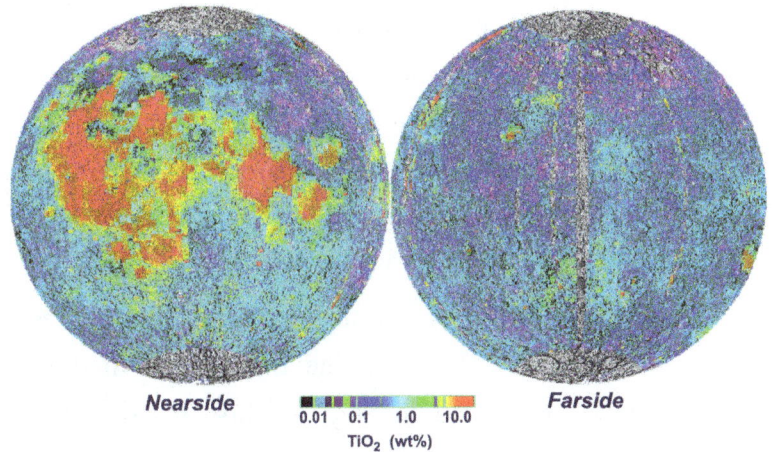

Nearside   0.01   0.1   1.0   10.0   *Farside*
TiO_2 (wt%)

Clementine Thorium Map

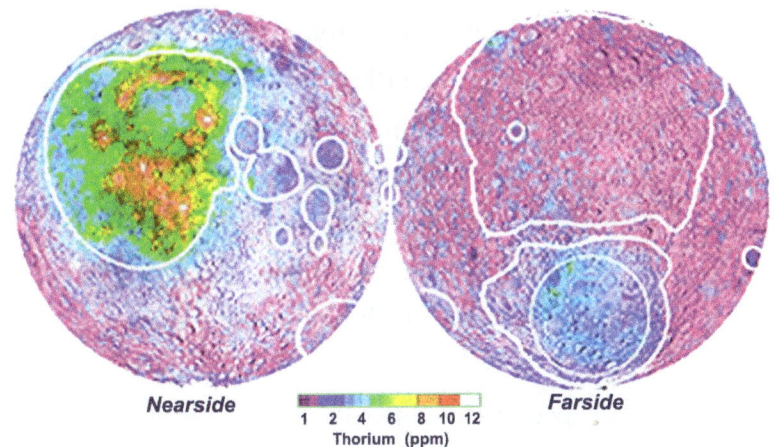

Nearside   1   2   4   6   8   10   12   *Farside*
Thorium (ppm)

What sets molten oxide electrolysis apart is its ability to directly electrolyze the regolith without supporting electrolyte in a one step process.

21   Larry A. Haskin, Department of Earth and Planetary Sciences and McDonnell Center for the Space Sciences, Washington University, St. Louis, MO 63130

Carbothermic Reduction

Carbothermic Reduction uses carbon to reduce metal oxides (ores) to produce metals such as iron in blast furnaces and metallurgical grade silicon. It is a three step process but the advantage it has over electrolysis is it can be operated at virtually any location on the moon with approximately the same efficiency.

☑ **Step 1. Minerals containing metallic oxides (iron, silicon, titanium, etc.) are reduced by reaction with methane to form carbon monoxide and hydrogen.**

☑ **Step 2. Carbon monoxide produced in Step 1 is reduced by reaction with hydrogen to form methane and water. The methane product is cycled back to use in Step 1.**

☑ **Step 3. Water formed in Step 2 is electrolyzed to form oxygen and hydrogen. The hydrogen is then cycled back to use in Step 2.**

The hydrogen formed in Steps 1 and 3 is consumed in Step 2. The methane formed in Step 2 is reused in Step 1. This closed cyclic process does not depend upon the presence of water or water precursors in the lunar materials. It will produce oxygen from ilmenite and silicate minerals in the lunar regolith regardless of their precise composition. Orbitec Inc. partnered with Physical Sciences Inc. and Kennedy Space Center to develop the technology.[22]

Radiance Technology proposes using plasma blasting instead of shipping explosives to the Moon.[23]

These and many other proposals require testing on the Moon before we can say which we will depend on. I cannot repeat too often, we are extremely lucky to have such a close neighbor as the Moon.

Two Places at Once

On May 31, 2011, a teleoperated construction machine accidentally punctured an oxygen cylinder at the Fukushima Dai-ichi nuclear plant causing the cylinder to explode.[24] The grapple-equipped excavator was clearing radioactive debris from the south side of the No. 4 reactor building when the accident happened. Cameras and other sensors mounted on the vehicle allow human operators miles away to clear debris in this hazardous environment. Strictly speaking, the machine is not a robot but a teleoperated machine.

http://www.seaeye.com

Teleoperating a machine is not new. The oil industry does it every day at the bottom of the ocean. The fact is, building anything in space or on the Moon will be impossible without using teleoperations and semiautonomous robotics. Relying on an astronaut to don a spacesuit and space walk to provide the labor in such an endeavor would be prohibitively expensive. We can create machines to be our hands, to be our eyes, to do all the things we would do if we were there but without all the life-support baggage.

22 Lunar and Planetary Institute, Los Alamos National Laboratory
23 8/5/2009: Pulsed Powered Plasma Blasting for Lunar Materials Processing

24 6/3/2011: Robotic Construction Machine Causes Explosion at Fukushima, ieee spectrum,

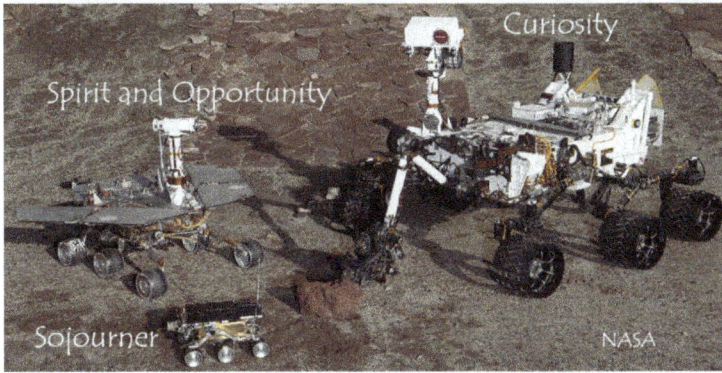

Humanity has already operated rovers on the Moon and Mars. The Russians were first sending two rovers to the Moon in the early 1970s. The technology back then allowed them to download high resolution images at an astounding rate of 21 seconds per frame. The two NASA rovers on Mars, Spirit and Opportunity, prove we can remote operate on a planet with a time delay up to twenty minutes long. NASAs newest rover, Curiosity, is on the Red Planet right now.

The latest NASA robot has it's own website.[25] Called Robonauts, they are dexterous humanoid robots with hands almost as capable as yours. In fact, Robonaut R2 is approaching human dexterity not only in the hands, but shoulders and neck as well. In NASA-speak, the advanced technology in R2 includes: optimized overlapping

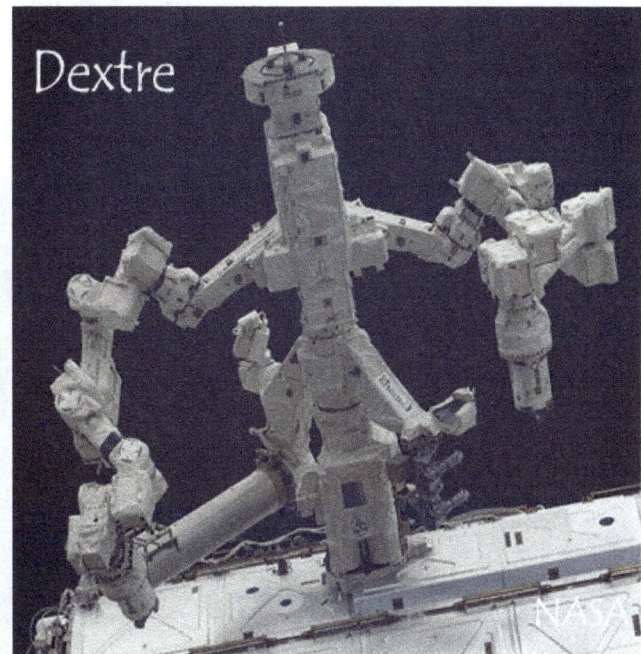

25 http://robonaut.jsc.nasa.gov/

dual arm dexterous workspace, series elastic joint technology, extended finger and thumb travel, miniaturized 6-axis load cells, redundant force sensing, ultra-high speed joint controllers, extreme neck travel, and high resolution camera and IR systems. What this boils down to is that R2 can use the same tools that a person would use. It removes the need for specialized tools just for robots. Robonauts will allow us to colonize space and build Powersats and never leave the ground.

The Special Purpose Dexterous Manipulator (Dextre) has been aboard the International Space Station since March 2008. It is a headless robot with two nine-foot arms made by the Canadian

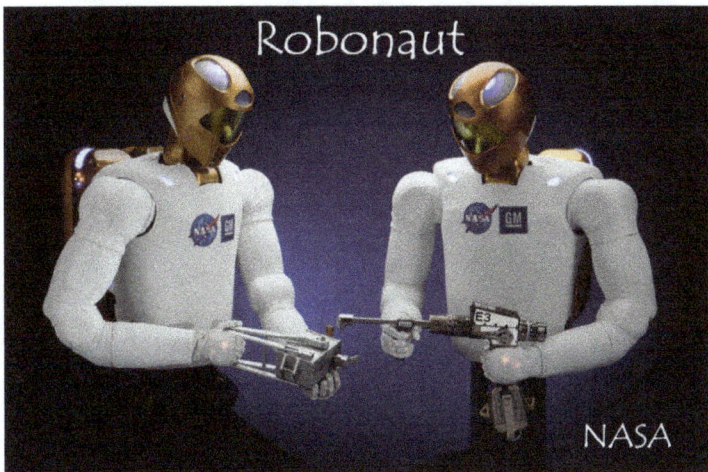

Space Agency.

After years of development and testing on the ISS, Dextre completed its first teleoperated job in February 2011. While the crew slept, controllers at NASA's Johnson Space Center in Houston, Texas, remotely dispatched Dextre to pluck two critical pieces of equipment from the hold of Japan's Kounotori 2 spacecraft.

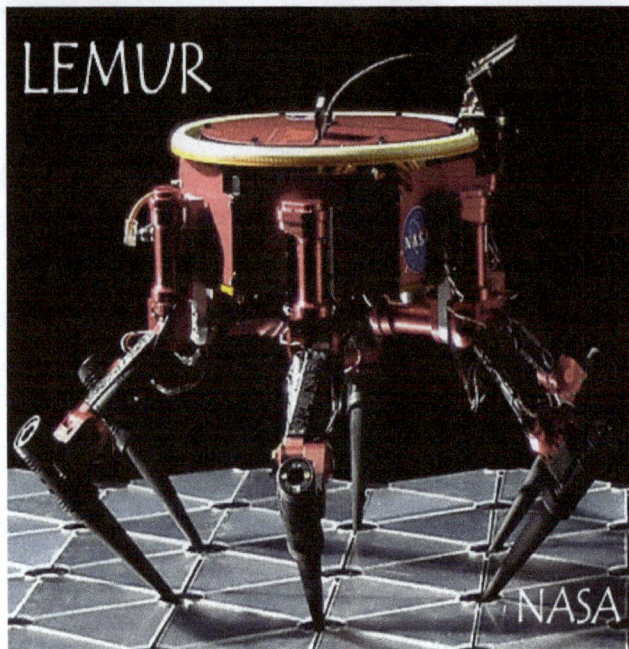
LEMUR
NASA

Developed by the Jet Propulsion Laboratory, the Limbed Excursion Mechanical Utility Robot (LEMUR) is a small, agile six-legged robot designed to work on orbital structures and spacecraft. It has a stereo camera allowing vision in any direction, and its "quick-connect" modular design allows tools to be changed quickly.

Robots should work together. Sample Return Rover (SRR) and its twin, SRR2K, are two technology prototype rovers testing software that has them cooperating while working on the Moon's surface. The software enables one person to control a team of robots or better yet, a team of humans controlling a team of robots.

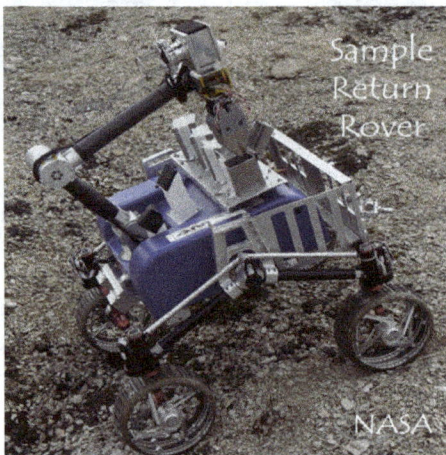
Sample Return Rover
NASA

We will always need to have a look ourselves, but until then, the Remote Surface Inspection system will have to do. This little rover is comprised of three subsystems: robot manipulation, graphical user interface, and multi-sensor inspection. Central to the latter are two color cameras and illuminators for visual inspection and a suite of sensors to detect temperature, gas vapors, eddy-currents, proximity, and force.

Developed by the University of Maryland, Ranger is a satellite servicing system providing real-time, model-based collision detection and avoidance for its two mechanical arms. This gives our teleoperated machines the ability to recognize and prevent us from puncturing something we shouldn't have while we are fixing the satellite. We will soon have the ability to repair the expensive satellites we place in orbit.

PREDATOR

Unmanned Aerial Vehicles (UAV) have become a major part of our global security strategy and are the next big thing in the sky over your head. We are all familiar with Predators. These are warplanes flown remotely by a Combat Systems Officer located anywhere in the world. Their use in Middle East conflicts is well documented and their list of kills is a long one.

Global Hawk

Northrop Grumman's Global Hawk is a spy plane capable of flying at 65,000 feet for 35 hours at 340 knots. It can image an area the size of the state of Illinois in just one mission.[26]

A-10 Warthog

In early 2012, the U.S. Defense Advanced Research Projects Agency (DARPA) added the latest in a long string of UAV projects when they announced the selection of Aurora Flight Sciences and Raytheon to develop an unmanned version of the A-10 Warthog, Close Air Support Aircraft and thereby, significantly ratcheting up UAV firepower.[27]

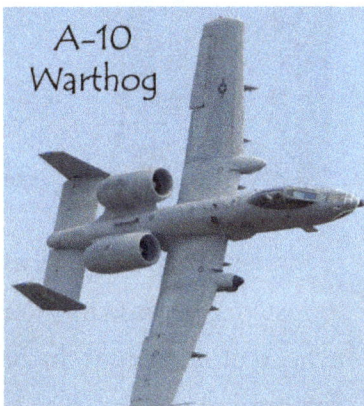

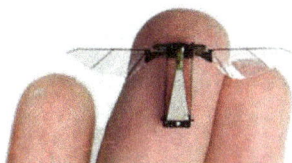

The smallest Unmanned Aerial Vehicle (UAV) is the size of a bug. It flies on gossamer wings and can perch virtually anywhere transmitting video and audio back to its master.

UAVs are the latest military technology transforming civilian society. The potential civilian market for UAVs will dwarf the military's use in the coming years. The hungriest market is the nation's 19,000 law enforcement agencies.[28] UAVs are currently deployed along the Mexican border with plans to expand.

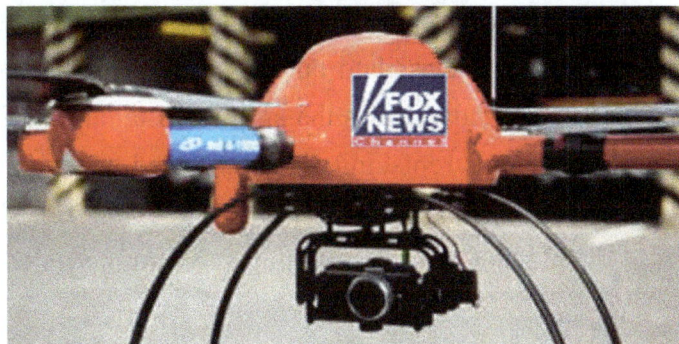

Rupert Murdock and News Corp[29] have already gotten into trouble using UAVs to spy on newsworthy subjects.

By taking the pilot out of the cockpit, warplanes and spy planes save weight which not only translates to more firepower, but increased mobility, extended duration, and miniaturization. Humans are large fragile creatures that get tired at the end of a ten hour mission or worse, pass out during a tight turn at the most inauspicious time. It becomes easy to switch out pilots when they are sitting in a simulator in the middle of a well protected base or police station. This is but a few examples of how we are learning to do things from a distance.

Every UAV, big and small, requires a robust and secure communications system to handle the complex operational requirements for fighting a warplane from a distance. It would be quite embarrassing to lose control of a UAV, especially if it is armed. Can you imagine criminals/terrorists hacking into the police/military network and taking over an armed UAV?

To alleviate that concern, on a clear January evening in 2012, a Delta IV rocket blasted off from Cape Canaveral carrying the Wideband Global SATCOM-4 satellite for the United States Air Force. WGS-4 is the fourth satellite in the militaries new high-capacity highly-encrypted

26   Northrop Grumman RQ-4 Block 10 Global Hawk
27   2/20/2012: Unmanned Version of A-10 On Way
28   2/28/2012: Civilian drone flights pushed in U.S. Joan Lowy, Associated Press
29   8/2/2011: FAA Looks Into News Corp's Daily Drone, Raising Questions About Who Gets To Fly Drones in The U.S.

satellite communications system.[30] The exact numbers are classified but this one satellite has as much bandwidth as the entire constellation of satellites it is replacing. Now we need to design one for the Moon.

Teleoperated planes, trains or automobiles

require a very sophisticated and secure communications network. Strictly speaking, it is not just about faster speeds. It is more about the bandwidth capacity available to each teleoperator. In late December 2010, the first High-Throughput Satellite (HTS) launched into geosynchronous orbit. The satellites total capacity is more than 70 gigabits per second (Gbps) which ranks it as the world's most powerful commercial spacecraft.[31] It is now providing internet service to most of Europe.

Video, photo sharing, games, voice over IP, and peer-to-peer networking across the internet is increasing commercial demand for satellite broadband worldwide. Right now, total global bandwidth doubles every two or three years with no end in sight. Using WildBlue residential service in the United States as an example, both times a new broadband satellite has come online, a spike in subscriber uptake followed and capacity sold out within months.[32] We gobble up

30 2/2/2012: Wideband Global SATCOM Satellite
31 2/16/2012: New-Generation of Broadband Satellites
32 WildBlue High Speed Internet and Exede Satellite Service

bandwidth as fast as they launch the satellites.

An HTS communications satellite placed at Lagrange Point L1 would service most of the nearside. Service would only be marginal out at the edges including the poles. A well-placed ground relay could make up the deficiency.

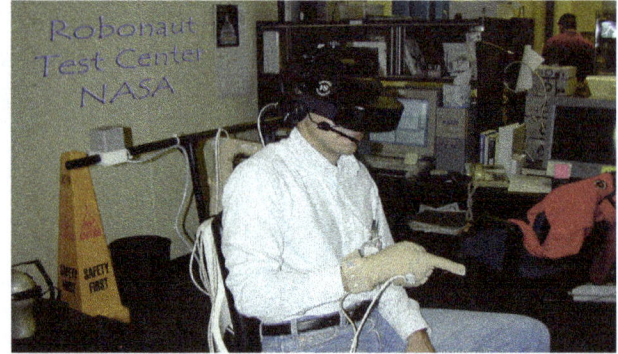

I can easily envision a time when hundreds of thousands, perhaps millions of people, teleoperate a host of machines in orbit and on the Moon performing construction tasks, mining, smelting, and working towards a world free of burning hydrocarbons or splitting atoms. Teleoperating will become a choice manufacturing job in the 21st century, at least three humans for every machine. Will America stand by and watch others do this work, or will we lead the world into a better tomorrow, Democracy (we the people) and Capitalism (the right to make a buck) walking hand in hand?

Staking Our Claim

Over the years, scientists and engineers have thoroughly considered where to put these orbital power satellites. The International Space Station circles the globe every 90 minutes or so which is good for testing and convincing the world that **SBSP** can work but it isn't practical for the bulk of the system. However, the Earth has a special orbit. A satellite placed in this orbit will go around Earth once a day. From the ground, it appears to

be as motionless as the tip of an enormous radio tower.

In other words, the satellite maintains the same position relative to the Earth's equatorial surface below it. This is geostationary or geosynchronous orbit (GEO) but many call it the Clarke Orbit in honor of the science fiction writer, Arthur C. Clarke, who first postulated using this orbit for communications satellites in 1945. Pointing a ground-based telescope up at an object in geostationary orbit, the satellite would appear to hover at the same point in the sky with respect to you. That's why the small dishes used to bring you satellite TV and internet access are not required to track their respective satellites once properly aligned.

Here's what you need to know about Earth's neighborhood. Low Earth Orbit is where the Space Station is. In fact, 200 miles up is still not out of the atmosphere. Any object in LEO gets hit by the occasional gas molecule slowing it down ever so slightly. This effect accumulates over time. Therefore, the station must periodically be boosted back into the proper orbit and the proper speed reestablished. If they didn't do this, the ISS would sink lower and lower and eventually, reenter the atmosphere in a great ball of fire. That may eventually be its fate regardless of our diligence.

The Clarke Orbit is one of a kind, literally. It is perfectly circular and lies in the Earth's

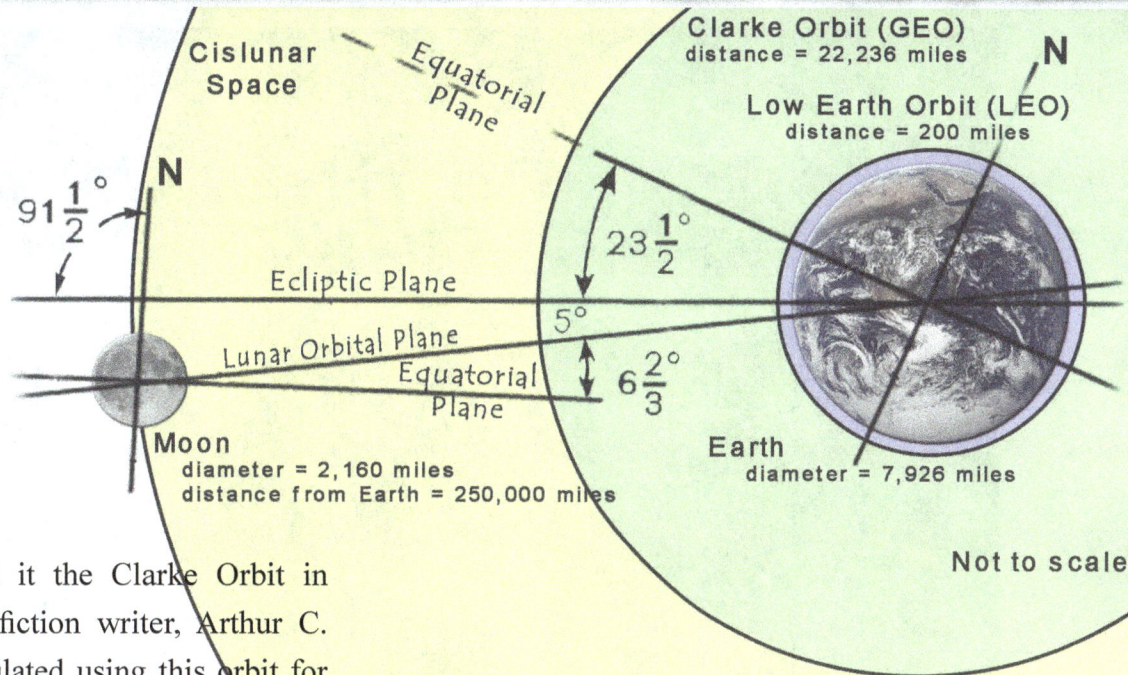

Equatorial Plane with a period equal to one Sidereal Day (23h 56m). The orbit has a radius of 26,199 miles (42,164 km) from the center of the Earth or 22,236 miles (35,786 km) above mean sea level. In 2002, the Clarke Orbit contained over 300 satellites. Talk about valuable real estate!

Between LEO and the Clarke Orbit is called Middle Earth Orbit. Catchy. I'll bet an engineer thought that one up. MEO is where the Global Positioning System satellite constellation operates as well as a host of specialized satellites ranging from military to weather observation.

Cislunar space is a loosely define term that encompasses all of that and more. Cislunar extends out at least to the Moon and even beyond. It is more accurately defined as anything in Earth orbit but with an emphasis on the lower orbits, from the Moon down.

A perfectly stable orbit is an ideal. In practice, satellites drift out of orbit because of the slightest perturbations such as the solar wind, radiation pressure, variations in the Earth's gravitational

Lagrange Points

field and the gravitational effect of Moon and Sun. Small forces will change things given enough time. So every once in a while, small thrusters are used to reestablish the orbit. This process is known as station-keeping. The VASIMR engine is perfectly suited for this job using hydrogen harvested on the Moon.

"The human race is remarkably fortunate in having so near at hand a full-sized world with which to experiment: before we aim for the planets, we will have had the chance of perfecting our astronautical techniques on our own satellite... the conquest of the Moon will be the necessary and inevitable prelude to remote and still more ambitious projects."

Arthur C. Clarke, 1951

Gravity interacts between two massive bodies, such as the Earth and Moon or the Earth and Sun, in mathematically predictable ways. The gravitational fields of the two bodies, combined with the centrifugal force of being in orbit, are in balance at five locations in space called the Lagrange Points after the mathematician who did the initial calculations. Think of them as gravitational eddies. Like a piece of flotsam on the river, a Lagrange Point will trap a third smaller body and cause it to remain stationary with respect to the massive bodies. All Lagrange Points lie on the Lunar Orbital Plane and have the same period as the Moon. L1, L2, and L3 are quasi-stable and require station keeping. L4 and L5 are stable regions that naturally entrap dust and other small bodies.

Lagrange Point L1 is the perfect location for humanity's next major space station. It is also the natural gateway to the Moon. This one-of-a-kind point in space is located about 210,000 miles from Earth on the line passing through the centers of both Earth and Moon. This puts it less than 40,000 miles above the geometric center of the Moon's nearside, the side that faces us.

The first space station at L1 will be a simple affair consisting of communications and global positioning hardware. However, it will not remain this way for long.

As we develop the Moon's resources, this station will grow as humans and their robotic partners transform it into a bustling manufacturing hub. It may one day contain a host of materials processing plants supporting large-scale lunar mining operations centered on Luna's first Mass Driver. And someday it will contain us.

The *Deep Space Climate Observatory* is a satellite proposed by Al Gore to occupy L1. In this position it will have a continuous full sunlight view of the Earth. The satellite's internet cameras will take a photograph every 15 minutes and share it with the world. We could watch the Earth as it passes through its monthly phases just as the Moon does when viewed from Earth. How would you like the ever changing view of Earth as your iPhone's wallpaper? Awesome! This also stakes our claim to the Lagrange Point.

Until something better comes along, L1 will serve as a fine jumping off point to the rest of the solar system. L1 is also the perfect location for the first practical Space Elevator down to the lunar surface. L1's foundry and solar furnace will forge the first geosynchronous **Space Based Solar Power** Satellite and other mega structures

that will eventually occupy L4 and L5.

We are not short on designs for these mega stations. A Bernel sphere will support 10 to 30 thousand colonists. The classic Stanford design is a torus capable of supporting 140 thousand, and Professor Gerard O'Neill favored a huge cylinder supporting millions. I suspect the final shape at L1 will be a compromise between a sphere and a torus thereby taking advantage of the meta-stable nature of the point to minimize station keeping. Any further speculation would turn this book into science fiction.

Energy security will be the motive and our military will lead the way into space but it will be civilian energy companies that will make **SBSP** the cash cow that drives a burgeoning cislunar economy. Tourism, both real and virtual, will also play a role. All of this will infuse a spirit of adventure that draws our young people to brave the vast inhospitable frontier of space.

One thing is certain, Lagrange points are unique, and if America does not move quickly, other nations will. After that, we can only hope to rent a room on their space stations. Take your pick, owner, renter, or outsider looking through the knothole. Pardon me, but I much prefer owner.

Stoking the Furnace

As far as the human species is concerned, the Sun has been around forever. The ancient Egyptians worshiped it. The ancient Greeks harnessed it. The Greeks learned that filling clear glass vases with water would focus sunlight. They called this crude lens a *burning glass*. As they got better at making lenses, the priests used it to light sacred fires and the doctors to cauterize

Four Solaire (solar furnace) at Odeillo

wounds. I wonder who was the first to fry ants with their new burning glass? Legend has it that Archimedes used burning lenses or mirrors or a combination of mirrors and lenses to ignite a fleet of Roman ships. Whatever he used, you can bet the Romans had their own when they came back. Their version of an arms race!

Like all technology, it wasn't just the Greeks and Romans who had solar powered lenses. It spread. Several 11th century hoards found at Viking sites near the city of Visby, Sweden[33], contained finely polished biconvex lenses made from high quality rock crystal. A few of the Visby lenses have an almost perfect elliptical shape, hard to match using modern equipment.[34]

The solar furnace itself isn't exactly new either. The ancient Greek word *heliocaminus* literally means solar furnace. It describes a glass-enclosed sunroom or hothouse that becomes hotter than the outside air temperature during the day. They were common throughout the Mediterranean.

Modern solar furnaces come in many shapes and sizes. A single small reflector can generate significant temperatures but we have learned to combine thousands of them together. Spain is a leader in developing ground based solar thermal power. The La Florida plant in Barcelona is the largest solar-thermo, electric-generating station in the world with an output over 50 MW.

The world's largest materials processing solar furnace was built in 1970 in the sunny Pyrenees Mountains on the French-Spanish border.[35] The Four Solaire at Odeillo uses a field of 10,000 mirrors bouncing sunlight onto a large concave mirror that in turn, focuses this enormous amount of incoming energy onto an area roughly the size of a basketball where temperatures can exceed 3,500 °C (6,330 °F).

Now it is time to take sunlight to a completely new level. A solar furnace on the surface of the Moon will have access to sunlight unfiltered by the atmosphere and in cislunar space unencumbered by that pesky night/day cycle.

Not only can orbital sunlight produce

---

33 Visby is the largest city on the island of Gotland, Sweden
34 1/19/2009: Viking Lenses from Visby, Sweden

35 12/29/2009: The Four Solaire Solar Furnace at Odeillo, France

Orbital Solar Furnace

electricity, sunlight has the power to sublimate[36] lunar ore and ionize[37] it, breaking it down to its most basic constituent parts. From there, all we need to do is find a way to separate the elements, perhaps the same way a mass spectrometer does, using electromagnetic fields.

Global Positioning System

According to the official government website, the Global Positioning System (GPS) is an American-taxpayer utility providing users with positioning, navigation, and timing services. GPS consists of Space, Control, and User segments. The Air Force manages the system to ensure the availability of at least 24 GPS satellites, 95% of the time.

Historically, the Space segment consisted of a core constellation of 24 operating satellites that transmit one-way signals giving that GPS satellite's current position and time. GPS satellites fly in Middle Earth Orbit (MEO) at an altitude of approximately 20,200 km. Each satellite circles

the Earth twice a day.

The satellites in the GPS core constellation surround the Earth in six equally spaced orbital planes, originally containing four satellites each. This 24-slot arrangement ensured at least four satellites were in view from virtually any point on the planet at all times.

In June 2011, the Air Force completed a GPS upgrade affectionately called the *Expandable 24*. They repositioned three of the original 24 satellites and added three more. As a result, GPS now effectively operates as a 27-satellite constellation. This improved coverage throughout the world.

The Air Force manages the Control segment of GPS. It consists of system wide monitoring and control stations that maintain the satellites in their proper orbits through occasional command maneuvers. They also adjust the satellite clocks when necessary, upload navigational data, and maintain the overall health and status of the satellite constellation. In short, they run the show.

The User segment has evolved into an

---

36 Changing from a solid to a gas without stopping at liquid
37 Remove one or more electrons thereby creating an electrical imbalance

incredible diversity of GPS receivers in use today. This is where the tire meets the road. A host of military and civilian applications, from cell phones, to long distance trucking, to missile guidance use GPS to know where they are anywhere on the globe to within a few feet. We have miniaturized the electronics and standardized the calculations until anyone can do it.

GPS is America's gift to the world. We don't charge a single penny for using its signals.[38] Any entrepreneur from any nation on the planet can incorporate GPS receivers into their products and charge customers like any good capitalist.

America isn't the only operator of a sophisticated navigational system. Russia has an operational global system called GLONASS and the European Union is developing Galileo, also global. Japan's navigational system is regional as well as India's proposed system, and China's current regional system is being upgraded to global.

Now we need to extend GPS to include all of cislunar space and specifically, the surface of the Moon. As a baseline, a lunar constellation might need as many satellites as the Earth system. Unless you are at the bottom of a crater or behind a mountain, a navigation beacon placed at the Lagrange Point L1 will always be visible from anywhere on the Moon's nearside. Conversely, Lagrange Point L2 will always be visible from anywhere on the Moon's farside. But in order to get an accurate reading, a receiver needs at least three satellites, more if you want finer resolution.

The design of a Lunar Positioning System (LPS) should follow the example of its earthly counterpart to ensure the maximum possible coverage even at the bottom of the darkest crater. It will be necessary if we are to return to the Moon and stay.

---

38 All about driving with GPS

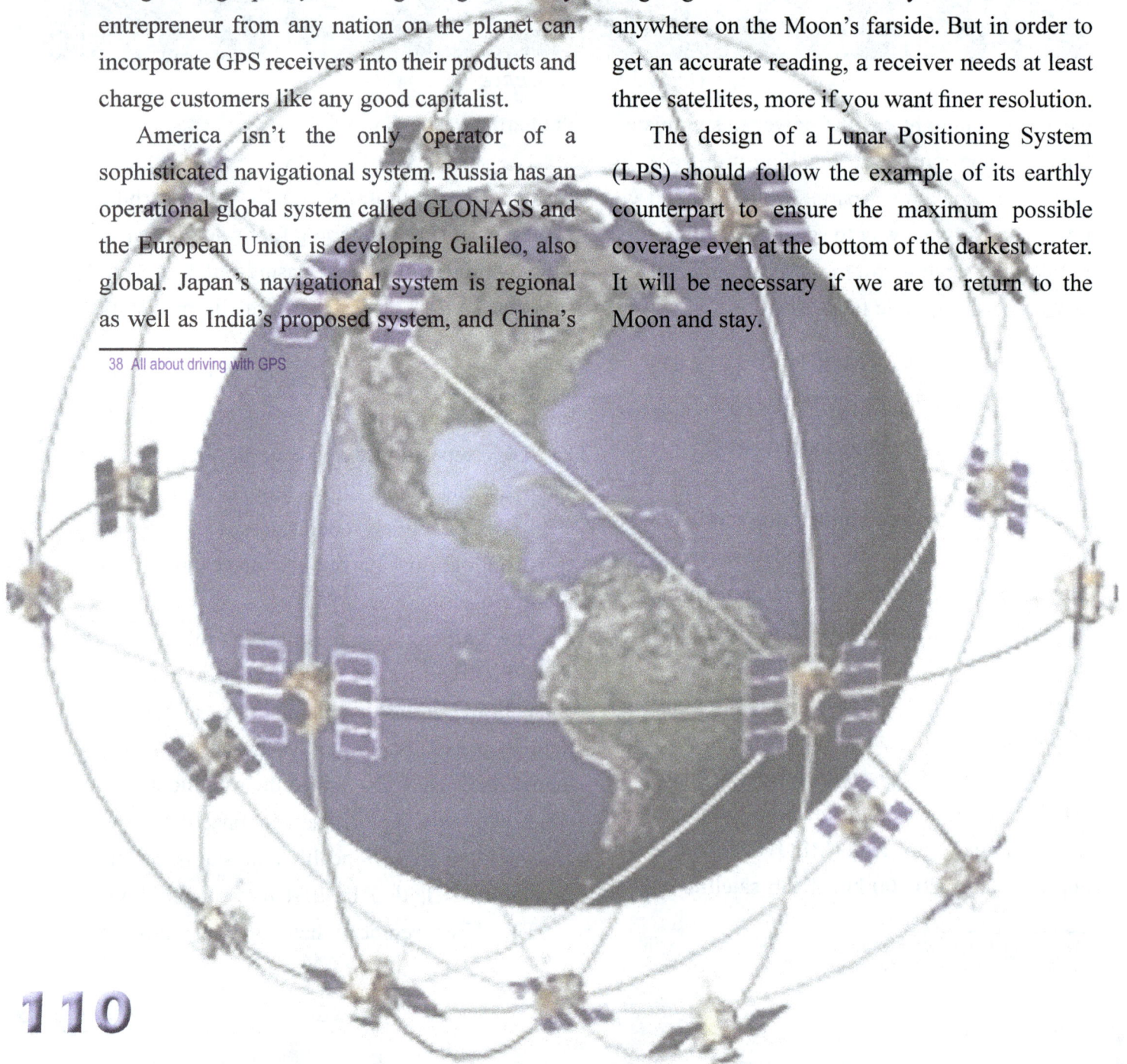

Conclusion and Summation

Where to go from here

Abundant cheap electricity is a key element in getting and maintaining high human living standards around the globe. Stated another way, electricity is the foundation of modern technology. Without it, we go back to sailing ships and the horse. Anyone who thinks for a moment that we could feed and clothe 7,000,000,000 people using only horsepower and sails, is nuts. Everything in your world depends on electricity either directly or indirectly. The food you eat, the clothes you wear, the car you drive, are all possible because of electricity. This cannot be overstated. Electricity is civilization.

Currently, the average electrical consumption per capita worldwide is about 2,850 kWh/year. Third world countries consume a fraction of that (390 kWh/year). Developing countries is less than half the average (1,250 kWh/year) while America is more than four times higher (13,750 kWh/year).[39] By working to improve the living standards around the globe, the total power demand increases. What once was the purview of Western Civilization is quickly becoming a worldwide phenomenon.

Wind turbines, geothermal, and ground based solar energy simply cannot support heavy industry. They will always have a place but cannot provide our planet's current baseload electrical consumption let alone the increased future demand as China, India and the rest of the world westernize. Nuclear energy is not the answer. The scale of building enough nuclear plants worldwide capable of meeting current demand is prohibitively expensive but perhaps more importantly, prohibitively dangerous on many levels. Burning coal, oil, and natural gas is polluting our environment and will eventually run out, maybe not tomorrow, but certainly within this century.

Space Based Solar Power is the only energy source capable of replacing coal, gas, and nuclear, and sustaining healthy growth into the future. **SBSP** is almost entirely free of greenhouse gas emissions, available at very large scales and deliverable anywhere on the planet. Using **SBSP**, it is possible to manufacture enough hydrogen to support a hydrogen-based economy. Using **SBSP**, it is possible to generate fresh water using energy-demanding desalination plants.

The positives far outweigh the negatives surrounding **SBSP** and the associated colonization effort. It would be in America's best interest to sponsor an immediate proof-of-concept demonstration project in full collaboration with our military and private industry. In the process, they would create a business model for the world to follow that will culminate in utility-grade **SBSP** electric power plants. It would not only be the birth of a new industry, but the opening of a new frontier. Considering the development timescales involved, the exponential population growth, and resource pressures we are currently experiencing, it is imperative that this work begins immediately, but just where do we start?

Many smart people want to believe that Tourism will be the driving force behind

39 U.S. Energy Information Administration, International Energy Outlook 2011

colonizing space. I have serious doubts. Others think the Free Market is the answer. Again, I don't think it can any more than it could have developed the Interstate Highway System. The only organization capable of a sustained colonization effort is government and here in America, that means either NASA or the military. My money (and yours) is on the military with a lot of help from NASA and private industry. The militaries enormous budget funds thousands of DARPA projects, past and present, which are paving the way to space. Electromagnetics, broadband communications, and unmanned teleoperated robotic systems are major elements in building a space based civilization. In the final analysis, it will be our men and women in uniform that lead us back to the Moon and solve the coming energy crisis.

At this time, in this political climate, after reviewing the available public data, it seems obvious that we are sending in the Marines!

Water

Mobility is the key to conquering space and developing **Space Based Solar Power**. Therefore, the first step in any long-range plan is harvesting the water ice we now know is at the lunar poles. With it, we can make the hydrogen-oxygen fuel to do many other things including maintenance, repair and expansion of our existing multibillion-dollar-a-year satellite industry.

Remote Operated System

Remote Operated Vehicles or ROVs have revolutionized our ability to work below the surface of the world's oceans. Whether the task is deep-sea scientific exploration, naval

operations or supporting offshore drilling, today's ROVs have evolved into powerful, well-instrumented, dexterous work platforms able to do virtually anything a man can do. Dextre, the NASA remotely operated robot onboard the ISS demonstrates the technology and gives us practical experience in orbit. The system NASA has designed lets a person on the ground link with Dextre and do a job in orbit. Teleoperating combat and spy planes has become commonplace and is posed to grow beyond its military birth. Soon every police force and news organization worth their salt will have their own UAV. We must now extend these capabilities to cislunar space and the Moon.

Workers on Earth, in LEO, or eventually, on Lagrange One space station, will operate an army of ROVs on the Moon and all across cislunar space including geosynchronous orbit where they will build a host of powersats. It will take decades to build enough to power the world. Internet cameras and sensors on virtually every machine will attract a global following of citizens joining in the fun, even if it is only as observers.

Perhaps teleoperating a prospector/explorer team on the Moon will be the centerpiece of some future reality show. It has got to be better than some of the shows currently on cable. I hope to live long enough to see the first Space Activist Network cable station begin broadcasting space news 24 hours a day. (*SAN brought to you by the National Space Society...*) I hope it is not as boring as the NASA channel.

Stake Our Claim

The Lagrange Points, all five of them, are unique, especially L1 and L2. Occupying these will be crucial to any hope of colonizing the Moon. Why not stake a claim to them by putting a small satellite in each? Even the simple solar powered internet camera that Al Gore suggested would do the job.

The National Space Society, the Moon Society and the host of space activists around the world could rally around this idea and even help fund the spacecraft. If a million activists from all across the globe donated $100 each, I believe we could put satellites in both L1 and L2. Even a few hundred thousand dollars would attract some cosponsors to share expenses. This is worth doing. One thing's clear, someone will be the first to occupy these points and it might as well be us.

Lunar Positioning System

Earth's Global Positioning System or GPS is a constellation of satellites. The lunar counterpart, or LPS, will have similar configuration but designed for the lunar environment, it may incorporate the Earth/Moon Lagrange Points to reduce the number of satellites needed for complete coverage. In general, Earth satellites have a ten-year life expectancy so the lunar LPS should be similar.

Lunar Communications Network

The same goes for the constellation of communications satellites, powered by the Sun and designed with a ten-year life. The network must support hundreds of thousands, perhaps millions, of simultaneous feeds between Earth and Luna. This will require a very robust satellite network to support the bandwidth requirements of complex robots such as Dextre and the generations of machines to follow.

One thing becomes instantly obvious, the Lagrange Points, L1, L2, L4, and L5, will take on a critical role in lunar communications and positioning systems. Sophisticated satellites capable of supporting LPS and carrying a heavy communications load should occupy the Lagrange Points early in the deployment. These locations will undoubtedly play a central role in the final design and greatly reduce the number of satellites needed.

NASA has sponsored many design studies over the years including one by Goddard's Communications, Standards and Technology Laboratory (CSTL). They have demonstrated a standards-based Internet Protocol communications system supporting lunar surface missions. The CSTL system incorporates wireless technologies, lunar surface communications base stations, lunar orbital relays, and Earth ground stations into one big happy network. We know how to do it.

Ice Rover

Once the network is in place, we can begin to send up the rovers. The initial rovers must be small and numerous. Built on a standard frame, they should allow for custom payloads ranging from ground penetrating radar to mineral prospecting. These rovers should be cheap and expendable and we should land them in clusters on the surface of the Moon. Everything is cheaper by the dozen. Some are worker rovers with payloads designed to support specific construction projects. Others are explorers or prospectors looking for the next mother lode of ice or titanium. University students will design some payloads for purely intellectual reasons and industrial engineers will

design their rovers for practical reasons, but all rovers will have many things in common such as engines, power source and communications. This will allow salvaging of parts to extend the useful life.

Once the ice has been located, miner rovers will harvest it. Miners will evolve into many specialized forms. Ice will be the first. All rovers will have cameras linked into the internet and LPS tracking capability so we will always know where they are.

Ice Processing Machines

Separating ice into its constituent parts will take several specialized machines. Solar power will supply the energy, and the almost continuous sunlight at a lunar pole is a big advantage. However, the ice is where the Sun doesn't shine so we must send miner rovers down into the bottom of some deep craters to collect the ice-rich ore. The rover must then deliver it to another machine back up where the Sun does shine for processing. First, we separate the ice from the rock and produce liquid water. Then we use electrolysis to separate the water into hydrogen and oxygen gas. The last step is to turn the gases into liquid. That involves pressure and coldness. Perhaps a clever engineer can take advantage of the extreme temperatures between lunar night (-200°C) and day (130°C). Another specialized rover collects the cryogenic fluids and delivers them to a lunar tanker.

Lunar Landers

Lunar tankers are the supply ships and traveling gas stations of the colonization effort. They are capable of landing payloads on the Moon, filling up their tanks with water, hydrogen, and oxygen

and launching from the Moon and back into orbit many times. Once in lunar orbit, they can deliver fuel anywhere in cislunar space. Using powerful chemical engines, they can quickly match orbits with satellite servicing vehicles, cislunar tugs, other landers, geosynchronous power satellites, or space stations bringing fuel where it is needed most.

Another type of lander is strictly cargo and never actually touches the Moon's surface. At an altitude of several hundred feet, the cargo starts lowering on a tether towards the Moon's surface as the lander simultaneously increases thrust. Just prior to impact, the tether releases in a coordinated move and gently places the cargo on the surface. This is called a skip-landing. The tether stays with the skip-lander and goes back into lunar orbit for another load.

Cislunar Tug

A variety of rockets delivers cargo to Low Earth Orbit but getting it from there to where it is going requires a specialized spacecraft called a Cislunar Tug. Powered by a VASAMR electromagnetic thruster running on lunar hydrogen, a cislunar tub takes longer but can deliver heavy payloads far more efficiently than a chemical powered rocket engine can.

Satellite Service Vehicle

Earth orbiting communications, resource, and military satellites are an annual multibillion-dollar industry. Being able to fix a damaged satellite or simply to fill up its station-keeping fuel tanks will be a lucrative activity using lunar fuel. VASIMR thrusters will power the Satellite Service Vehicle and onboard robots with the dexterity of Dextre will perform complex tasks under remote human guidance. Adaptability will be the hallmark of space technology. For instance, when a cislunar tug is not available, the service vehicle can pick up a new satellite in LEO and deliver it to geosynchronous orbit.

Adaptability requires standardizing critical components throughout the system. Modular construction techniques and specialized robotics will also be essential. Hardware flexibility will be important when disaster strikes. Murphy's Law – if anything can go wrong, it will.

Metals

Getting hydrogen and oxygen from water is a piece of cake compared to extracting metals. Moreover, if this endeavor is to work, we must become very good Moon miners fast. Again, the bulk of the work done on the Moon, collecting ore, processing it, and smelting it into usable shapes, will take significantly more effort, most of which will be done remotely.

Regolith Miner

The low fruit on the tree is what passes for soil on the Moon. Regolith is a mixture of pulverized Moon material and the objects that did the pulverizing. The thickness of the layer and the composition of the regolith vary from place to place. Without an atmosphere, the surface of the Moon has been ground into particles of various sizes, all with broken-glass sharp edges, nasty stuff if you were to breathe it in. Machines must account for the tiny dust sized particles getting into places where it shouldn't; bearings, switches, seals, anything with moving parts.

However, since the first step in processing ore is pulverizing it, dealing with regolith is to our advantage. Therefore, our miners must be able to scoop up the regolith, store it onboard, and deliver it to the appropriate processor.

Metals Processing Machines

These machines will come in a wide variety designed to harvest the iron, titanium and aluminum used to build many things in space and on the Moon. Of immediate need will be expanding our footprint on the Moon.

The mass of the processing machines that come all the way from Earth must be kept at a minimum. Therefore, sending small machines to build bigger machines from lunar metals will save money and create a self sustaining outpost sooner rather than later. In the beginning, we ship only the critical components of a bigger miner to the Moon and build the frame, skin and storage bins from metals extracted from the Moon. Once we have the capability to build one vehicle, we can build a hundred of any shape or size. The same goes for the processing machines themselves. Start with small machines designed to create bigger machines.

Solar Furnace

Sunlight is the gift that keeps on giving. The largest solar furnace in the world is in France. It can easily melt iron. We need to adapt this technology in both cislunar space and on the surface of the Moon. A solar furnace is a critical step in obtaining metals and other prized elements from lunar regolith and it requires only sunlight to run.

The space station located at Lagrange Point L1 will eventually have its own solar furnace, taking advantage of the zero-g environment of space to forge a variety of alloys and form them into parts. We may even design a furnace and its

associated assembly line to manufacture the solar cells used on our powersats. The more we can do with lunar resources, the better.

Mass Driver

Getting processed metal and other cargo off the Moon will be much more efficient using a mass driver instead of burning fuel. In the past, a mass driver seemed out of our reach. O'Neil's vision was of a kilometers-long machine that sucked huge amounts of power. Today that has changed. The latest generation of naval Railgun has the muzzle velocity to power its projectiles into low lunar orbit (LLO) or even past escape velocity and is the size of a semitrailer. The one we design for the Moon will be modular and smaller, you can bet on it. I imagine it will still take a lot of electrical power but we can deal with that.

Lagrange One

The first full-scale space based solar power satellite will not be in geosynchronous orbit about Earth. It will be in Lagrange Point L1. Mining and processing activity all across the Moon's nearside needs power to function throughout the two-week night. The easiest way to do it is with a powersat in L1. On the Moon we needn't worry about atmospheric absorption or in keeping the power density low. Mirrors reflecting and focusing sunlight down to the Moon's surface sounds easy right now but high-power lasers could also do the job.

As the system grows, we will need to transfer energy around cislunar space to fully take advantage of a **SBSP** system and the easiest way to do that is with lasers. The Japanese are developing the technology to do this. Lasers will be another aspect of **SBSP** that has possible military connotations but it is not alone. There are many things about space colonization that will have both military and civilian aspects. There simply isn't any way to avoid this and I don't think that's a real problem at this point. Any real push into space must, by necessity, include our military. It will not succeed without them.

Lagrange One and Two will take on increasingly important functions as the colonization project proceeds. L1, and later L2, will be the target of mass driver projectiles, but by the time they arrive they will have virtually zero velocity. Envision throwing a rock up into the air and just as it reaches its greatest height, someone reaches out and grabs it.

It is critical to take possession of Lagrange One and Lagrange Two. We can share but if China beats us to them, we will always be second best in the eyes of the world. National pride is not the only thing at stake here. The economic benefit of their possession is impossible to calculate. It is like Queen Isabella's advisor asking what the economic benefits were for investing three ships in his crazy idea as he handed Columbus a bunch of flags to take with him.

Power Satellites

Development and space testing of the orbital elements that will make up the Power Satellites or powersats, has already started. The Japan Aerospace Exploration Agency, or JAXA, is moving forward on their plan to demonstrate the first powersat by the end of the decade and a full **SBSP** system by 2030. Ambitious. Now it is time for America to do the same.

The first powersats will direct their energy to places it is most needed such as disasters and military sites. The ease at which the ground receiving antennas are put up, proving that the microwave beam is harmless, and last but not least, delivering power into the grid will create a huge market for **Space Based Solar Power**. Only after the infrastructure on the Moon has matured can we begin building utility grade solar power satellites in the quantities required. And once we start, it will take many years to complete. Similar in nature to the American Interstate Highway System, work on the Global Energy Service may never end. New powersats, just like new roads, might be added each year. Service and repair on such a huge system provides humanity with job security for the 21st Century.

Only when we are deep into the plan will people follow the machines and physically colonize the Moon. However, when we finally do join them, we will do so in force. The New Frontier will be as fierce as the Wild West ever thought of being. Those with a short attention span or lack of detail will not last long. If the vacuum doesn't get you, a meteor will. As Robert Heinlein wrote, *The Moon Is a Harsh Mistress* and space is even worse.

The stations at Lagrange One and Two will undergo many transformations over time including becoming human habitats long before we permanently return to the Moon ourselves. Teleoperating machines down on the surface of the Moon from them will be just a little faster, a little more responsive, during critical operations.

People

This is only a rough outline of what is needed to colonize our Moon and create a **Space Based Solar Power** civilization. With it come many challenges but also great reward. We must convince our business leaders that there is money to be made in space and our politicians that it is worth a national investment in time and treasure. Our military has already staked their claim to the high ground and now it is time for us to let them take it.

Other Books

Shadow on the Moon

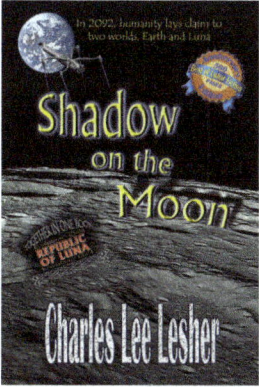

The Republic of Luna is humanities first extraterrestrial nation. Science, genetics and a humanistic society mark it as a target for the powerful Islamic Brotherhood, a global empire with billions of believers. Luna is a world created by pioneers whose only religion is the humane treatment of one another in their common struggle to survive the ultimate hostile environment, space. The heroes that conquered the moon must now defend it. Thankfully, they have a few tricks up their sleeve. Shadow on the Moon combines *Evolution's Child - Earthman, Evolution's Child - Lunarian, Revelation's Child*, and *Science of the Republic* into one 500 page Anniversary Print Edition.

ISBN: 978-0-977723-56-0 Paperback

Aldrin Station - Rise of Luna

Aldrin Station is a collection of short stories illuminating Lunarian history from the dawn of mankind to its expansion into space and colonizing the moon. These are stories of the families and individuals that play a role in establishing the Republic of Luna.

ISBN 978-1-938586-05-7 Paperback
ISBN 978-1-938586-00-2 eBook

Science of the Republic

A collection of articles, maps, and tables that help the reader understand the science and technology of the Republic.

ISBN 978-1-938586-09-5 Paperback
ISBN 978-1-938586-04-0 eBook

Evolution's Child - Earthman

Book One: Lazarus Sheffield is a man without a planet by the time he meets Lindsey on his way to Heaven's Gate Space Station. Lindsey quickly determines that the nervous guy sitting next to her is a high ranking government official on the run from one of history's most repressive governments, the totalitarian theocracy otherwise known as the North American Federation. She decides to help him and introduces Lazarus to some of Luna's finest citizens. So begins Book One of Shadow on the Moon.

ISBN 978-1-938586-06-4 Paperback
ISBN 978-1-938586-01-9 eBook

Evolution's Child - Lunarian

Book Two: Tempel Dugan leads a group of Lunarians against impossible odds. They call themselves Quan Kiai. These young warriors, and a few more like them, are all that stands between the Republic of Luna and total annihilation but things are not always as they seem.

ISBN 978-1-938586-07-1 Paperback
ISBN 978-1-938586-02-6 eBook

Revelation's Child

Book Three: Quan Kiai and the other Lunarian warriors have their backs against the wall. Fight or die. They fight! They fight in their great underground cities, they fight cross the surface of the moon, and they fight in orbital space. Earth and Luna become locked in humanities first interplanetary war, one that history will call the First Lunarian War.

ISBN 978-1-938586-08-8 Paperback
ISBN 978-1-938586-03-3 eBook

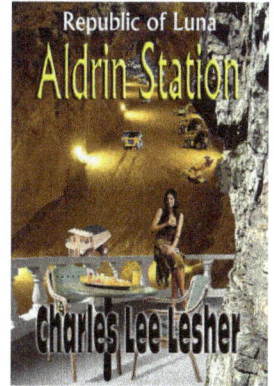

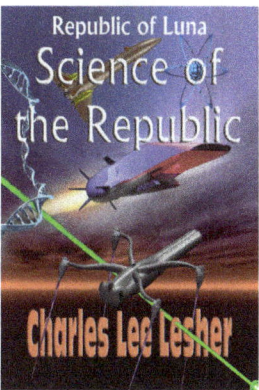

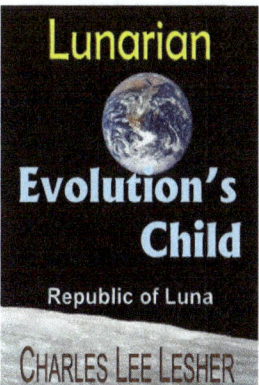

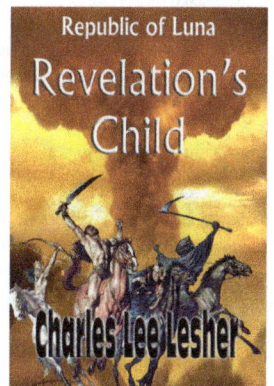

Index

www.ingramcontent.com/pod-product-compliance
Lightning Source LLC
Chambersburg PA
CBHW051337200326
41519CB00026B/7457